DATE DUE

4-18-02	ILS 783085-9		

Amino Acid Analysis Protocols

METHODS IN MOLECULAR BIOLOGY™

John M. Walker, SERIES EDITOR

METHODS IN MOLECULAR BIOLOGY™

Amino Acid Analysis Protocols

Edited by

Catherine Cooper

Nicolle Packer

and

Keith Williams

Proteome Systems Ltd, North Ryde, Australia

Humana Press ✳ Totowa, New Jersey

QD431.25.A53 A45 2001

Amino acid analysis
protocols /

The content and opinions expressed in this book are the sole work of the authors and editors, who have warranted due diligence in the creation and issuance of their work. The publisher, editors, and authors are not responsible for errors or omissions or for any consequences arising from the information or opinions presented in this book and make no warranty, express or implied, with respect to its contents.

This publication is printed on acid-free paper. ∞
ANSI Z39.48-1984 (American Standards Institute) Permanence of Paper for Printed Library Materials.

Cover design by Patricia F. Cleary.

For additional copies, pricing for bulk purchases, and/or information about other Humana titles, contact Humana at the above address or at any of the following numbers: Tel: 973-256-1699; Fax: 973-256-8341; E-mail: humana@humanapr.com, or visit our Website at www.humanapress.com

Photocopy Authorization Policy:

Printed in the United States of America. 10 9 8 7 6 5 4 3 2 1

Library of Congress Cataloging-in-Publication Data

Amino acid analysis protocols / edited by Catherine Cooper, Nicolle Packer and Keith Williams.
        p. cm. -- (Methods in molecular biology ; 159)
    Includes bibliographical references and index.
    ISBN 0-89603-656-1
    1. Amino acids--Analysis. I. Cooper, Catherine, 1970- II. Packer, Nicolle. III.
Williams, Keith, 1948- IV. Methods in molecular biology (Totowa, N.J.) ; v. 159.

    QD431.25.A53 A45 2000
    547'.75046--dc21

                                                    00-023798

Preface

Amino acid analysis is a technique that has become commonplace in biotechnology, biomedical, and food analysis laboratories. This book describes a variety of amino acid analysis techniques and how each technique can be used to answer specific biological questions.

The first two chapters in *Amino Acid Analysis Protocols* introduce the concepts, basic theory, and practice of amino acid analysis. The following chapters give detailed instructions on various methods and their applications.

As highlighted, there are many different approaches to amino acid analysis, but in all cases the results depend heavily on the quality of the sample. Therefore a new way to desalt samples prior to hydrolysis is covered as an introductory chapter (Chapter 3), and most authors have devoted a section to sample preparation, especially to the collection and storage of bodily fluids.

Some of the amino acid analysis methods described in this book are based on HPLC separation and analysis after precolumn derivatization. The precolumn derivatization techniques described use (a) 6-aminoquinolyl-*N*-hydroxy-succinimidyl carbamate (AQC) (Chapters 4 and 8); (b) 1-fluoro-2,4-dinitrophenyl-5-L-alanine amide (Marfey's reagent), which allows separation and analysis of enantiomeric amino acids (Chapter 5); (c) *O*-phthalaldehyde (OPA) (Chapters 6 and 10); (d) butylisothiocyanate (BITC) and benzylisothiocyante (BZITC) (Chapter 11); (e) phenylisothiocyanate (PITC) (Chapters 12 and 13); (f) ammonium-7-fluorobenzo-2-oxa-1,3-diazole-4-sulfonate (SBD-F) (Chapter 17); and (g) 9-fluorenylmethyl-chloroformate (FMOC-Cl) (Chapter 10).

Techniques have been described in which gas chromatography is used to separate and analyze (a) amino acids after *N*(*O,S*)-isoBOC methyl ester derivatization (Chapter 9); (b) *N*-isoBOC methyl esters of *O*-phosphoamino acids (Chapter 14); and (c) *N*(*S*)-isopropoxycarbonyl methyl esters derivatives of sulfur amino acids, glutathione, and other related aminothiols such as CysGly (Chapter 15). New techniques based on capillary electrophoresis separation (Chapter 16), high-performance anion-exchange chromatography (Chapter 7), and mass spectrometry of isotopically labeled proteins (Chapter 18) are also presented.

The applications of amino acid analysis are extremely varied and the technique remains the best means of accurate protein quantitation. Examples given in *Amino Acid Analysis Protocols* include the use of amino acid analysis for identification of picomolar amounts of protein on PVDF membranes (Chapter 8). The measurements of amino acids in bodily fluids and tissues such as urine (Chapters 9, 12, 14, 15), blood (Chapters 9, 10, 12, 14, 15, 17), seminal plasma (Chapter 6), or skeletal muscle tissue (Chapter 16), and measurement in the presence of high lipid content, such as in porcine lung (Chapter 13), are useful to help to identify diseases associated with changes in amino acid metabolism. Amino acid analysis, for example, is important to the study of such disorders as maple syrup urine disease (accumulation of branched-chain L-amino acids), phenylketonuria (high concentrations of phenylalanine), atherosclerosis (elevated levels of homocysteine), and galactosemia (often high concentrations of methonine). Amino acid and glucose analysis in fermentation broths of cell cultures (Chapter 7) enables the development of a feeding strategy that maintains the correct levels of nutrients. This is important since the use of such systems to make recombinant products is increasing. A method to determine the amino acid composition of foods (Chapter 11) is also included.

In addition to the standard methods used to separate the 20 commonly occurring amino acids, the analysis of unusual and modified amino acids is also addressed. Specifically, the analysis of homocysteine for monitoring the development of atherosclerosis (Chapter 17); hydroxyproline, a major amino acid found in collagen (Chapter 16); phosphoamino acids, which are difficult because they are acid labile (Chapter 14); aminothiols, such as cysteinylglycine and cystathionine (Chapter 15); and glycated lysine, implicated in diabetic complications and Alzheimer's disease (Chapter 18).

Overall *Amino Acid Analysis Protocols* presents an up-to-date, detailed methodology reference for a broad range of current techniques being used for amino acid analysis.

Catherine Cooper
Nicolle Packer
Keith Williams

Contents

Contributors

NEBOJSA AVDALOVIC • *Research and Development, Dionex Corporation, Sunnyvale, CA*

VICTOR BARRETO • *Research and Development, Dionex Corporation, Sunnyvale, CA*

PHILIPP CHRISTEN, MD • *Biochemisches Institut der Universität Zürich, Zürich, Switzerland*

QINGYI CHU • *Doheny Eye Institute, Los Angeles, CA*

STEVEN A. COHEN • *Waters Corporation, Milford, MA*

CATHERINE COOPER • *Proteome Systems Ltd, North Ryde, Australia*

NANCY DENSLOW • *Protein Chemistry Core Laboratory, Interdisciplinary Center for Biotechnology Research (ICBR), University of Florida, Gainesville, FL*

ISABELLA FERMO • *Laboratory of Separative Techniques, Department of Laboratory Medicine, Scientific Institute H. S. Raffaele and School of Medicine, Milan, Italy*

C. DAVID FORSTER • *School of Biomedical Sciences, Nottingham University Medical School, Queen's Medical Centre, Nottingham, UK*

TSUTOMU FUJIMURA • *Division of Biochemical Analysis, Central Laboratory of Medical Sciences, Juntendo University School of Medicine, Tokyo, Japan*

REED J. HARRIS • *Analytical Chemistry Department, Genentech Inc., San Francisco, CA*

PETR JANDIK • *Research and Development, Dionex Corporation, Sunnyvale, CA*

JAN JOHANSSON • *Department of Medical Biochemistry and Biophysics, Karolinska Institutet, Stockholm, Sweden*

HIROYUKI KATAOKA • *Faculty of Pharmaceutical Sciences, Okayama University, Okayama, Japan*

SAIKO KAZUNO • *Division of Biochemical Analysis, Central Laboratory of Medical Sciences, Juntendo University School of Medicine, Tokyo, Japan*

RODNEY G. KECK • *Analytical Chemistry Department, Genentech Inc., San Francisco, CA*

SUNIL KOCHHAR • *Department of Bioscience, Nestlé Research Center, Lausanne, Switzerland*

HUA LIU • *Department of Research, Texas Scottish Rite Hospital for Children, Dallas, TX*

FRANK D. MACCHI • *Analytical Chemistry Department, Genentech Inc., San Francisco, CA*

MASAMI MAKITA • *Faculty of Pharmaceutical Sciences, Okayama University, Okayama, Japan*

CHARLES A. MARSDEN • *School of Biomedical Sciences, Nottingham University Medical School, Queen's Medical Centre, Nottingham, UK*

SAYURI MATSUMURA • *Faculty of Pharmaceutical Sciences, Okayama University, Okayama, Japan*

BARBARA MOURATOU • *Biochemisches Institut der Universität Zürich, Zürich, Switzerland*

KIMIE MURAYAMA • *Division of Biochemical Analysis, Central Laboratory of Medical Sciences, Juntendo University School of Medicine, Tokyo, Japan*

KIYOHIKO NAKAI • *Faculty of Pharmaceutical Sciences, Okayama University, Okayama, Japan*

BERYL J. ORTWERTH • *Mason Eye Institute, University of Missouri, Columbia, MO*

NICOLLE PACKER • *Proteome Systems Ltd, North Ryde, Australia*

RITA PARONI • *Laboratory of Separative Techniques, Department of Laboratory Medicine, Scientific Institute H. S. Raffaele and School of Medicine, Milan, Italy*

CHRISTOPHER POHL • *Research and Development, Dionex Corporation, Sunnyvale, CA*

MALLADI PRABHAKARAM • *Mason Eye Institute, University of Missouri, Columbia, MO*

NORIHISA SAKIYAMA • *Faculty of Pharmaceutical Sciences, Okayama University, Okayama, Japan*

FELICITY J. SHEN • *Analytical Chemistry Department, Genentech Inc., San Francisco, CA*

ROY A. SHERWOOD • *Department of Clinical Biochemistry, King's College Hospital, London, UK*

NORIKO SHINDO • *Division of Biochemical Analysis, Central Laboratory of Medical Sciences, Juntendo University School of Medicine, Tokyo, Japan*

JEAN B. SMITH • *Department of Chemistry, University of Nebraska, Lincoln, NE*

MARGARETA STARK • *Department of Medical Biochemistry and Biophysics, Karolinska Institutet, Stockholm, Sweden*

KIYOMI TAKAGI • *Faculty of Pharmaceutical Sciences, Okayama University, Okayama, Japan*

HIROFUMI TANAKA • *Faculty of Pharmaceutical Sciences, Okayama University, Okayama, Japan*

MARGARET I. TYLER • *Australian Proteome Analysis Facility, Macquarie University, Sydney, Australia*

YUKIZO UENO • *Faculty of Pharmaceutical Sciences, Okayama University, Okayama, Japan*

KEITH WILLIAMS • *Proteome Systems Ltd, North Ryde, Australia*

KANG-LYUNG WOO • *Division of Life Science, Kyungnam University, Masan, South Korea*

SHIGEO YAMAMOTO • *Faculty of Pharmaceutical Sciences, Okayama University, Okayama, Japan*

MICHAEL ZEECE • *Department of Food Science, University of Nebraska, Lincoln, NE*

LI ZHANG • *Protein Chemistry Core Laboratory, Interdisciplinary Center for Biotechnology Research (ICBR), University of Florida, Gainesville, FL*

1

Amino Acid Analysis

An Overview

Margaret I. Tyler

1. Importance and Utility

Amino acids are found either in the free state or as linear chains in peptides and proteins. There are 20 commonly occurring amino acids in proteins, which are shown in **Table 1**. Amino acid analysis has an important role in the study of the composition of proteins, foods, and feedstuffs. Free amino acids are also determined in biological material, such as plasma and urine, and in fruit juice and wine. When it is performed on a pure protein, amino acid analysis is capable of identifying the protein (*2,3*, and Chapter 8 in this volume), and the analysis is also used as a prerequisite for Edman degradation and mass spectrometry and to determine the most suitable enzymatic or chemical digestion method for further study of the protein. It is also a useful method for quantitating the amount of protein in a sample (*see* Chapter 2 in this volume) and can give more accurate results than colorimetric methods.

2. Historical View

The earliest experiments on the acid hydrolysis of proteins were performed by Braconnot in 1820, in which concentrated sulphuric acid was used to hydrolyze gelatin, wool, and muscle fibers (*4*). Various reagents for performing protein hydrolysis were tried over the next 100 years, with 6 *M* HCl becoming the most widely accepted reagent. In 1972, Moore and Stein (*5*) were awarded the Nobel Prize for developing an automated instrument for separation of amino acids on an ion-exchange resin and quantitation of them using ninhydrin.

More recently, high-performance liquid chromatographs (HPLCs) have been configured for amino acid analysis. Some methods use postcolumn deriv-

From: *Methods in Molecular Biology*, vol. 159: *Amino Acid Analysis Protocols*
Edited by: C. Cooper, N. Packer, and K. Williams © Humana Press Inc., Totowa, NJ

Table 1
Common Amino Acids

Name	Symbol 3 letter	Symbol 1 letter	Essential for humans (1)
Acidic amino acids			
Aspartic acid	Asp	D	No
Glutamic acid	Glu	E	No
Neutral amino acids			
Alanine	Ala	A	No
Asparagine	Asn	N	No
Cysteine	Cys	C	No
Glutamine	Gln	Q	No
Glycine	Gly	G	No
Isoleucine	Ile	I	Yes
Leucine	Leu	L	Yes
Methionine	Met	M	Yes
Phenylalanine	Phe	F	Yes
Serine	Ser	S	No
Threonine	Thr	T	Yes
Tryptophan	Trp	W	Yes
Tyrosine	Tyr	Y	No
Valine	Val	V	Yes
Basic amino acids			
Arginine	Arg	R	Yes
Histidine	His	H	Yes
Lysine	Lys	K	Yes
Imino acid			
Proline	Pro	P	No

atization in which the amino acids are separated on an ion-exchange column followed by derivatization with ninhydrin (*6*, and Chapter 2 in this volume), fluorescamine (*7*), or o-phthalaldehyde (*8*). Another approach has been to derivatize amino acids prior to separation on a reversed-phase HPLC column. Examples of this technique are dansyl (*9*), phenylisothiocyanate (PITC) (*10*, and Chapters 12 and 13 of this volume), 9-fluorenylmethyl chloroformate (Fmoc) (*11*), and 6-aminoquinolyl-N-hydroxysuccinimyl carbamate (AQC) (*12*, and Chapters 4 and 8 in this volume).

3. Sensitivity

Amino acid analysis can be performed accurately at the fmol level by methods employing fluorescence detection, whereas for derivatives detected by

Table 2
Comparison of Different Derivatization Chemistries
for Amino Acid Analysis

	Ninhydrin	OPA	OPA	PITC	Fmoc	AQC
Derivatization type[a]	postc	postc	prec	prec	prec	prec
Detection mode[b]	c	f	f	UV	f	f
Sensitivity	pmol	fmol	fmol	pmol	fmol	fmol
Chromatography[c]	i.e.	i.e.	r.p.	r.p.	r.p.	r.p

OPA, orthophthalaldehyde; PITC, phenylisothiocyanate; Fmoc, 9-fluorenylmethyl chloroformate; AQC, 6-aminoquinolyl-N-hydroxysuccinimidyl carbamate.
[a]postc, postcolumn; prec, precolumn.
[b]c, colorimetry; f, fluorescence; UV, ultraviolet.
[c]i.e, ion exchange; r.p., reversed-phase HPLC.

ultraviolet (UV) light, the analysis is at the pmol level. **Table 2** gives a comparison of the various derivatization chemistries and their sensitivities. Annual studies comparing the various methods have been carried out by the Association of Biomolecular Resource Facilities (ABRF) *(13,14)*. Strydom and Cohen *(15)* have compared AQC and PITC chemistries and found AQC derivatives to be more stable.

4. Difficult Amino Acids

4.1. Tryptophan

Tryptophan is destroyed in acid hydrolysis. Alkaline hydrolysis with NaOH, Ba (OH)$_2$, or LiOH have been used particularly in the hydrolysis of food and feedstuffs *(16,17)*. However, acid hydrolysis is still needed to determine the other amino acids.

There have been a number of methods published for the determination of tryptophan that use the standard 6 *M* HCl hydrolysis in the presence of additives, some of which include thioglycolic acid *(18)*, beta-mercapto ethanol *(19)*, and mercaptoethanesulfonic acid *(20)*.

4.2. Cysteine and Cystine

Cysteine and cystine are unstable during acid hydrolysis, particularly in the presence of carbohydrate. The total content of cysteine and cystine can be determined by oxidizing the protein with performic acid, which converts both forms to cysteic acid and methionine to methionine sulphone. The protein is then hydrolyzed with 6 *M* HCl *(17)*.

Disulphide compounds such as dithiopropionic acid and dithiobutyric acid have been proposed as protecting agents for cysteine and cystine during acid hydrolysis *(21)*. The use of dithiodiglycolic acid during the acid hydrolysis step, followed by phenylisothiocyanate derivatization, allows cysteine and cystine plus all common hydrolysate amino acids (excluding tryptophan) to be determined *(22)*.

Reduction of disulphide bridges, followed by alkylation of cysteine with iodoacetic acid or 4-vinylpyridine is also used to determine the cysteine-plus-cystine content of proteins *(23)*. Alkylation with acrylamide produces cysteine-S-propionamide, which is converted to cysteine-S-propionic acid during acid hydrolysis *(24)*.

4.3. Asparagine and Glutamine

Asparagine and glutamine are amide derivatives of aspartic acid and glutamic acid, respectively. During acid hydrolysis, which cleaves amide bonds, asparagine is converted to aspartic acid and glutamine to glutamic acid. Thus, the amount determined for aspartic acid represents the total of aspartic acid and asparagine and similarly for glutamic acid and glutamine.

4.4. D amino Acids

The D-amino acid content of a protein or peptide can be determined by employing a short partial acid hydrolysis, followed by an enzymatic hydrolysis with pronase, and then with leucine aminopeptidase and peptidyl-D-amino acid hydrolase *(25)*.

5. Modified Amino Acids

Phosphorylated amino acids are able to be analyzed using a variety of different chemistries *(26)*, but the ABRF 1993 study found that precolumn methods were more successful *(27)*. Phosphoserine *(28,29)*, phosphothreonine *(29)*, and phosphotyrosine *(29)* have varying stabilities. Highest recoveries for phosphoserine and phosphotyrosine are produced with hydrolysis time of 60 min or less at 110°C, whereas for phosphothreonine, a hydrolysis time of 2 h gave better results *(26)*. Chapter 14 in this volume covers the analysis of phosphoamino acids more extensively.

There are many other rarer amino acids and derivatives that can be analyzed. These include hydroxyproline *(17,30*, and Chapter 16 in this volume) and hydoxylysine *(17,30*, and Chapter 2 of this volume), found in collagen. Taurine has dietary importance and can be readily determined in infant formulas, pet food, plasma, urine, and tissue extracts *(17*, and Chapter 10 of this volume). Posttranslational modifications, including glycosylated amino acids *(31*, and Chapters 2 and 7 of this volume) and glycated amino acids *(32,33)*, are important in studying protein function. Chapter 18 of this volume describes the application of mass spectrometry to the analysis of glycated amino acids.

6. Limitations—Contaminants and Precautions

The accuracy of amino acid analysis is very dependent on the integrity of the sample. Cleanliness of all surfaces the sample contacts is essential, as is the purity of all reagents used. Traces of salts, metals, or detergents can effect the accuracy of results.

The hydrolysis step is particularly important, as was demonstrated in the ABRF 1994 AAA collaborative study *(34)*. Many laboratories now satisfactorily perform a 1-h hydrolysis in 6 *M* HCl at 150°C under vacuum. The traditional method uses 6 *M* HCl for 20–24 h at 110°C under vacuum. Losses of serine, threonine, and to a lesser extent, tyrosine may occur under these conditions. During acid hydrolysis, some amide bonds between aliphatic amino acids are more difficult to cleave. The Ala–Ala, Ile–Ile, Val–Val, Val–Ile, Ile–Val, and Ala–Val linkages are resistant to hydrolysis and may need a longer hydrolysis time of 48 or 72 h at 110°C *(35)*.

References

1. *Encyclopaedia of Food Science Food Technology and Nutrition*, vol. 1 (Macrae, R., Robinson, R. K., and Sadler, M. J., eds.), Academic, London, p. 149.
2. Hobohm, U., Houthaeve, T., and Sander, C. (1994) Amino acid analysis and protein database compositional search as a rapid and inexpensive method to identify proteins. *Anal. Biochem.* **222,** 202–209.
3. Schegg, K. M., Denslow, N. D., Andersen, T. T., Bao, Y. A., Cohen, S. A., Mahrenholz, A. M., and Mann, K. (1997) Quantitation and identification of proteins by amino acid analysis: ABRF-96 collaborative trial, in *Techniques in Protein Chemistry VIII* (Marshak, D., ed.), Academic, San Diego, CA, pp. 207–216.
4. Braconnot, H. (1820) *Ann. Chim. Phys.* **13,** 113.
5. Moore, S. and Stein, W. H. (1963) Chromatographic determination of amino acids by the use of automatic recording equipment, in *Methods in Enzymology*, vol. 6 (Colowick, S. P. and Kaplan, N. O., eds.), Academic, New York, pp. 819–831.
6. Samejima, K., Dairman, W., and Udenfriend, S. (1971) Condensation of ninhydrin with aldehydes and primary amines to yield highly fluorescent ternary products. 1. Studies on the mechanism of the reaction and some characteristics of the condensation product. *Anal. Biochem.* **42,** 222–236.
7. Stein, S., Bohlen, P., Stone, J., Dairman, W., and Udenfriend, S. (1973) Amino acid analysis with fluorescamine at the picomole level. *Arch. Biochem. Biophys.* **155,** 202–212.
8. Roth, M. (1971) Fluorescence reaction for amino acids. *Anal. Chem.* **43,** 880–882.
9. Tapuhi, Y., Schmidt, D. E., Lindner, W., and Karger, B. L. (1981) Dansylation of amino acids for high-performance liquid chromatography analysis. *Anal. Biochem.* **115,** 123–129.
10. Bidlingmeyer, B. A., Cohen, S. A., and Tarvin, T. (1984) Rapid analysis of amino acids using pre-column derivatisation. *J. Chromatog.* **336,** 93–104.

11. Haynes, P. A., Sheumack, D., Kibby, J., and Redmond, J. W. (1991) Amino acid analysis using derivatisation with 9-fluorenylmethyl chloroformate and reversed-phase high performance liquid chromatography. *J. Chromatog.* **540,** 177–185.

12. Strydom, D. J. and Cohen, S. A. (1993) in *Techniques in Protein Chemistry IV* (Angeletti, R. H., ed.), Academic, San Diego, CA, pp. 299–307.

13. Marenholz, A. M., Denslow, N. D., Andersen, T. T., Schegg, K. M., Mann, K., Cohen, S. A., et al. (1996) Amino acid analysis — recovery from PVDF membranes: ABRF-95AAA collaborative trial, in *Techniques in Protein Chemistry VII* (Marsak, D. R., ed.), Academic, San Diego, CA, pp. 323–330.

14. Tarr, G. E., Paxton, R. J., Pan, Y. C.-E, Ericsson, L. H., and Crabb, J. W. (1991) Amino acid analysis 1990: the third collaborative study from the association of biomolecular resource facilities (ABRF) in *Techniques in Protein Chemistry II* (Villafranca, J. J., ed.), Academic, San Diego, CA, pp. 139–150.

15. Strydom, D. J. and Cohen, S. A. (1994) Comparison of amino acid analyses by phenylisothiocyanate and 6-aminoquinolyl-N-hydroxysuccinimyl carbamate precolumn derivatisation. *Anal. Biochem.* **222,** 19–28.

16. Delhaye, S. and Landry, J. (1986) High-performance liquid chromatography and ultraviolet spectrophotometry for quantitation of tryptophan in barytic hydrolysates. *Anal. Biochem.* **159,** 175–178.

17. Cohen, S. A., Meys, M., and Tarvin, T. L. (1988) *The PicoTag Method. A Manual of Advanced Techniques for Amino Acid Analysis.* Waters Chromatography Division, Millipore Corp., Milford, MA.

18. Yokote, Y., Murayama, A., and Akahane, K. (1985) Recovery of tryptophan from 25-minute acid hydrolysates of protein. *Anal. Biochem.* **152,** 245–249.

19. Ng, L. T., Pascaud, A., and Pascaud, M. (1987) Hydrochloric acid hydrolysis of proteins and determination of tryptophan by reversed-phase high-performance liquid chromatography. *Anal. Biochem.* **167,** 47–52.

20. Yamada, H, Moriya, H., and Tsugita, A. (1991) Development of an acid hydrolysis method with high recoveries of tryptophan and cysteine for microquantities of protein. *Anal. Biochem.* **198,** 1–5.

21. Barkholt, V. and Jensen, A. L. (1989) Amino acid analysis: determination of cysteine plus half-cystine in proteins after hydrochloric acid hydrolysis with a disulphide compound as additive. *Anal. Biochem.* **177,** 318–322.

22. Hoogerheide, J. G. and Campbell, C. M. (1992) Determination of cysteine plus half-cystine in protein and peptide hydrolysates: use of dithiodiglycolic acid and phenylisothiocyanate derivatisation. *Anal. Biochem.* **201,** 146–151.

23. Inglis, A. S. (1983) Single hydrolysis method for all amino acids, including cysteine and tryptophan, in *Methods in Enzymology,* vol. 91. Academic, San Diego, CA, pp. 26–36.

24. Yan, J. X., Kett, W. C., Herbert, B. R., Gooley, A. A., Packer, N. H., and Williams, K. L. (1998) Identification and quantitation of cysteine in proteins separated by gel electrophoresis. *J. Chromatog.* **813,** 187–200.

25. D'Aniello, A., Petrucelli, L., Gardner, C., and Fisher, G. (1993) Improved method for hydrolysing proteins and peptides without introducing racemization and for determining their true D-amino acid content. *Anal. Biochem.* **213,** 290–295.
26. Yan, J. X., Packer, N. H., Gooley, A. A., and Williams, K. L. (1998) Protein phosphorylation: technologies for the identification of phosphoamino acids. *J. Chromatog. A.* **808,** 23–41.
27. Yüksel, K. Ü., Andersen, T. T., Apostol, I., Fox, J. W., Crabb, J. W., Paxton, R. J., and Strydom, D. J. (1994) Amino acid analysis of phospho-peptides: ABRF-93AAA, in *Techniques in Protein Chemistry V* (Crabb, J. W., ed.), Academic, San Diego, CA, pp. 231–240.
28. Meyer, H. E., Swiderek, K., Hoffmann-Posorske, E., Korte, H., and Heilmeyer, L. M., Jr. (1987) Quantitative determination of phosphoserine by high-performance liquid chromatography as the phenylthiocarbamyl-S-ethylcysteine. Application to picomolar amounts of peptides and proteins. *J. Chromatog.* **397,** 113–121.
29. Ringer, D. P. (1991) Separation of phosphotyrosine, phosphoserine and phosphothreonine by high-performance liquid chromatography, in *Methods in Enzymology*, vol. 201. Academic, San Diego, CA, pp. 3–10.
30. Waters AccQ.Tag Amino Acid Analysis System Operators Manual (1993). Millipore Corp., Melford, MA.
31. Packer, N. H., Lawson, M. A., Jardine, D. R., Sanchez, J. C., and Gooley, A. A. (1998) Analyzing glycoproteins separated by two-dimensional gel electrophoresis. *Electrophoresis* **19,** 981–988.
32. Walton, D. J. and McPherson, J. D. (1987) Analysis of glycated amino acids by high-performance liquid chromatography of phenylthiocarbamyl derivatives. *Anal. Biochem.* **164,** 547–553.
33. Cayot, P. and Tainturier, G. (1997) The quantitation of protein amino groups by the trinitrobenzenesulfonic acid method: a reexamination. *Anal. Biochem.* **249,** 184–200.
34. Yüksel, K. Ü., Andersen, T. T., Apostol, I., Fox, J. W., Crabb, J. W., Paxton, R. J., and Strydom, D. J. (1994) The hydrolysis process and the quality of amino acid analysis: ABRF-94AAA collaborative trial, in *Techniques in Protein Chemistry VI* (Crabb, J. W., ed.), Academic, San Diego, CA, pp. 185–192.
35. Ozols, J. (1990) Amino acid analysis, in *Methods in Enzymology*, vol. 182. Academic, San Diego, CA, pp. 587–601.

2

Amino Acid Analysis, Using Postcolumn Ninhydrin Detection, in a Biotechnology Laboratory

Frank D. Macchi, Felicity J. Shen, Rodney G. Keck, and Reed J. Harris

1. Introduction

Although lacking the speed and sensitivity of more widely heralded techniques such as mass spectrometry, amino acid analysis remains an indispensable tool in a complete biotechnology laboratory responsible for the analysis of protein pharmaceuticals.

Moore and Stein developed the first automated amino acid analyzer, combining cation–exchange chromatographic separation of amino acids with postcolumn ninhydrin detection (1). Commercial instruments based on this design were introduced in the early 1960s, though many manufacturers have abandoned this technology in favor of precolumn amino acid derivatization with separations based on reversed-phase chromatography (2–4) (see **Note 1**).

In our product development role, we still rely on amino acid analysis to generate key quantitative and qualitative data. Amino acid analysis after acid hydrolysis remains the best method for absolute protein/peptide quantitation, limited in accuracy and precision only by sample handling. We produced an Excel macro to process these data; the macro transfers and converts the amino acid molar quantities into useful values such as composition (residues per mol) and concentration. In addition, we employ several specialized amino acid analysis applications to monitor structural aspects of some of our recombinant products.

De novo biosynthesis of leucine in bacteria will lead to a minor amount of norleucine (Nle) production (5), particularly if recombinant proteins are produced in fermentations that have been depleted of leucine (6). The side-chain of Nle (-CH_2-CH_2-CH_2-CH_3) is similar enough to methionine (-CH_2-CH_2-S-CH_3) that some of the tRNA[Met] will be acylated by Nle, leading to incorporation of Nle at

From: *Methods in Molecular Biology*, vol. 159: *Amino Acid Analysis Protocols*
Edited by: C. Cooper, N. Packer, and K. Williams © Humana Press Inc., Totowa, NJ

Met positions *(6,7)*. When this occurs, Nle may be incorporated at a low level at every Met position, and amino acid analysis is often the only method able to detect this substitution.

Hydroxylysine (Hyl) is a common modification of lysine residues found at -*Lys*-Gly- positions in collagens and collagen-like domains of modular proteins *(8)*. This modification is also found at certain solvent-accessible -*Lys*-Gly- sites in noncollagenous proteins, usually at substoichiometric levels *(9)*. Amino acid analysis is a useful screening technique for the identification of Hyl-containing recombinant proteins produced by mammalian cells.

The analysis of recombinant proteins using carboxypeptidases may still be required to assign the C-terminus when the polypeptide chain is extensively modified, thus ruling out making a C-terminal assignment based solely on mass and N-terminal analyses, or in cases where the C-terminal peptide cannot be assigned in a peptide map. When carboxypeptidase analyses are needed, a modified amino acid analysis program is needed to resolve Gln and Asn (which are not found in acid hydrolysates) from other amino acids.

Assignment of Asn-linked glycosylation sites is greatly facilitated by prior knowledge of the -*Asn*-Xaa-Thr/Ser/Cys- consensus sequence sites *(10)*, and specific endoglycosidases, such as peptide:N-glycosidase F can be employed to quantitatively release all known types of Asn-linked oligosaccharides *(11)*. O-linked sites are harder to assign, as these are found in less-stringent sequence motifs *(12–14)*, and there is no universal endoglycosidase for O-glycans except for endo-α-*N*-acetylgalactosaminidase, which can only release the disaccharide Gal(β1→3)GalNAc. In addition, O-glycosylation is often substoichiometric.

In mammalian cell products, at least two N-acetylglucosamine (GlcNAc) residues are found in Asn-linked oligosaccharides, whereas N-acetylgalactosamine (GalNAc) is found at the reducing terminus of the most common (mucin-type) O-linked oligosaccharides. A cation–exchange-based amino acid analyzer can easily be modified for the analysis of the amino sugars glucosamine ($GlcNH_2$) and galactosamine ($GalNH_2$) from acid hydrolysis of GlcNAc and GalNAc, respectively, allowing confirmation of the presence of most oligosaccharide types. In glycoproteins, HPLC fractions from peptide digests can be screened using amino sugar analysis to identify glycopeptides for further analysis.

Regulated biotechnology products are usually tested for identity using HPLC maps after peptide digestion *(15,16)*. A key aspect of the digestion step for most proteins is obtaining complete reduction of all disulfide bonds, followed by complete alkylation of cysteines without the introduction of artifacts (e.g., methionine S-alkylation) *(17)*. Amino acid analysis can be used to monitor cysteine alkylation levels for reduced proteins, such as are obtained after alkylation with iodoacetic acid, iodoacetamide or 4-vinylpyridine.

2. Materials

2.1. Equipment

1. 1-mL hydrolysis ampoules (Bellco, Vineland, NJ; part number 4019-00001) (*see* **Note 2**).
2. Savant SpeedVac.
3. Oxygen/methane flame.
4. Glass knife (Bethlehem Apparatus Co., Hellertown, PA).
5. 1/4" ID × 5/8" OD Tygon tubing.
6. Model 6300 analyzers (Beckman Instruments, now Beckman Coulter, Fullerton CA). The sum of the 440 nm and 570 nm absorbances is converted to digital format using a PE Model 900 A/D converter, and the data are collected by a PE Turbochrom Model 4.1 data system (*see* **Notes 3–5**).
7. Lithium-exchange column (Beckman part number 338075, 4.6 × 200 mm).

2.2. Reagents and Solutions

1. Constant boiling (6 *N*) HCl ampoules are obtained from Pierce (Rockland, IL) (*see* **Note 6**).
2. Mobile phase buffers purchased from Beckman Instruments include sodium citrate buffers Na-D, Na-E, Na-F, Na-R, and Na-S; lithium citrate buffers include Li-A, Li-B, Li-C, Li-D, Li-R, and Li-S.
3. Ninhydrin kits (Nin-Rx) are also purchased from Beckman; these must be mixed thoroughly before use (usually 2 h at room temperature), and care must be taken to avoid skin discoloration because of contact with ninhydrin-containing materials.
4. Dialysis may be used to desalt samples into dilute acetic acid prepared from deionized water (Milli-Q, Millipore) and Mallinckrodt U.S.P. grade glacial acetic acid.
5. Amino acid standards: are diluted from the stock Beckman standard (part number 338088) with Na-S buffer to final concentration of 40 nmol/mL or 20 nmol/mL (*see* **Note 7**).
6. 2 *N* glacial acetic acid.

3. Methods

3.1. Sample Preparation

Proteins should be desalted to obtain optimal compositional data. Dialysis against 0.1% acetic acid removes salts while keeping proteins in solution, but quantitative data will often require direct hydrolysis (i.e., without dilution or sample losses introduced during dialysis). When proteins must be analyzed without desalting, neutral buffers such as 50 m*M* Tris can be used without compromising the results. Excipients to avoid include urea (which generates abundant ammonia during hydrolysis), sugars (which caramelize during hydrolysis), and detergents such as the polysorbate and Triton types that can

Table 1
Standard Amino Acid Analysis

Time (min)	Event	Conditions
0.0	Sample injection	Na-E buffer, 48°C
8.5	Start temp. gradient	48°C to 60°C in 8 min
24.5	Buffer change	Na-E to Na-F
41.0	Buffer change	Na-F to Na-D
78.0	Reagent pump	Ninhydrin to water
79.0	Buffer change	Na-D to Na-R
80.0	Buffer change	Na-R to Na-E
82.5	Temperature change	60°C to 48°C
84.0	Reagent pump	Water to ninhydrin
97.0	Recyle (start next run)	

Buffer pump: 16 mL/h.
Reagent pump: 8 mL/h.

damage cation–exchange columns. Samples in enzyme-linked immunosorbent assays (ELISA)-type buffers should be avoided as they typically contain albumin or gelatins, whose amino acids cannot be distinguished after hydrolysis from the protein of interest. Peptides generally can be desalted by reverse phase (RP)-HPLC using volatile solvents such as 0.1% TFA in water/acetonitrile.

1. Place samples in hydrolysis ampoules (*see* **Note 8**), then dry under vacuum using a Savant SpeedVac.
2. Place approx 100 µL of 6 N HCl in the lower part of the ampoule (*see* **Note 9**). Freeze in a dry ice/ethanol bath, attached to a vacuum system via 1/4" ID × 5/8" OD Tygon tubing, then slowly thaw and evacuate to < 150 mtorr.
3. Use an oxygen/methane flame to seal the neck of the tube at the constriction.
4. Place the sealed ampoules in a 110°C oven for 24 h (*see* **Note 10**), then allow to cool before opening after scoring them with a glass knife.
5. Remove the acid by vacuum centrifugation, again using a Savant system, with a NaOH trap inserted between the centrifuge and cold trap.
6. After hydrolysis and acid removal, samples that contain 0.5–10 µg of protein, or 0.1–1 nmol of peptide fractions should be reconstituted with 60–200 µL of Na-S sample buffer (*see* **Note 11**).

3.2. Protein/Peptide Quantitation

1. Subject triplicate samples containing 0.5–10 µg of protein or 0.1–1 nmol of peptide to 24-h hydrolysis *in vacuo* as aforementioned.
2. Follow the standard operating conditions given in **Table 1** (*see* **Note 12**). A standard chromatogram containing 2 nmol of each component is shown in **Fig. 1**.

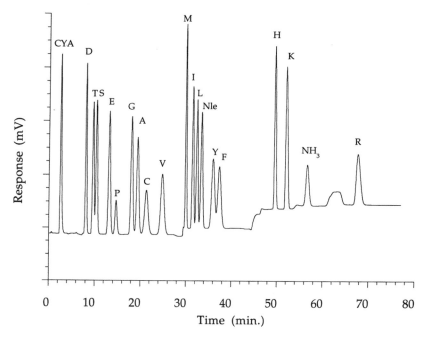

Fig. 1. Analysis of a standard amino acid mixture. The standard contains 2 nmol of each component except for NH_3. Operating parameters are given in **Table 1**.

3. Peak area data from the Turbochrom system are converted to nmol values by external standard calibration; internal standards are not necessary if a reliable autosampler is used.

4. The amino acid nmol values are also automatically converted to .tx0 files that can be imported into a custom Microsoft Excel program called the AAA MACRO (**Table 2**) for analysis using a PC-based computer.

5. The first step in running the AAA MACRO is to open a template, such as the example "protein.xls" given in **Table 3**. The residues per mol and molecular mass calculations must be modified and saved for each different protein/peptide; Asn and Asp are reported as Asx, whereas Gln, Glu and pyroglutamate are reported as Glx.

6. The macro asks for some background information (e.g., requestor's name, sample name, number of replicates), sample prep information (e.g., volumes of hydrolysate loaded vs reconstitution volume, original sample volume), then processes the data, providing a single-page report showing calculated compositions and concentration, as shown in **Fig. 2** (*see* **Note 13–15**).

3.3. Norleucine Incorporation

1. Detection of trace Nle levels in *Escherichia coli*-derived proteins require 24-h hydrolysis of 25–100 µg of protein (*see* **Note 16**).

Table 2
Amino Acid Analysis Data Conversion Macro

Commands	
Macro 4(a)	
=ACTIVATE("MACRO4A.XLM")	
=HIDE()	
=\YC4DATA	
=SELECT(!B4)	
=INPUT("Requestor's Name?",2,"Name","")	Asks for Requestor's Name
=IF(A7=FALSE,HALT())	Halts macro if Cancel button is clicked
=FORMULA(A7)	Returns name to data worksheet cell B4
=SELECT(!B5)	
=INPUT("Requestor's Extension?",1,"Telephone extension","")	Asks for Requestor's Extension
=IF(A11=FALSE,HALT())	Halts macro if Cancel button is clicked
=FORMULA(A11)	Returns extension to data worksheet cell B5
=SELECT(!E4)	
=INPUT("Sample to be analyzed?",2,"Sample name","")	Asks for protein to be analyzed
=IF(A15=FALSE,HALT())	Halts macro if Cancel button is clicked
=FORMULA(A15)	Returns protein's name to data worksheet cell E4
=SELECT(!E5)	
=INPUT("Requestor's Mail Stop?",1,"Mail Stop","")	Asks for Requestor's MailStop
=IF(A19=FALSE,HALT())	Halts macro if Cancel button is clicked
=FORMULA(A19)	Returns Requestor's MailStop to data worksheet cell E5
=SELECT(!B28)	
=INPUT("Molecular Mass of protein to be analyzed?",1, "Molecular Mass (g/mole)","")	Asks for MW
=IF(A23=FALSE,HALT())	Halts macro if Cancel button is clicked
=FORMULA(A23)	Returns MW to data worksheet cell B28
=SELECT(!F29)	
=INPUT("µL in Ampoule?",1," Ampoule volume (µL)","")	Asks for amount sample put in ampoule
=IF(A27=FALSE,HALT())	
=FORMULA(A27)	
=SELECT(!F30)	
=INPUT("Sample reconstitution volume?",1,"Reconstituted volume (µL)","")	Asks for reconstitution volume of sample
=IF(A31=FALSE,HALT())	Halts macro if Cancel button is clicked
=FORMULA(A31)	
=SELECT(!F31)	

Code	Comment
=INPUT("Dilution Factor?",1,"Dilution factor","")	Asks if samples were diluted prior to analysis
=IF(A35=FALSE,HALT())	Halts macro if Cancel button is clicked
=FORMULA(A35)	
=SELECT(!B1)	Selects cell B1 on worksheet
=INPUT("How many replicates will you be analyzing today?",1,"Number of replicates","")	Asks how many replicate samples will be processed
=IF(A39=FALSE,HALT())	Halts macro if Cancel button is clicked
=FORMULA(A39)	Places users sample number in cell B2
=SET.NAME("counter1",1)	Resets counter1
=SELECT(!B41)	Selects active cell to be B6
=DIRECTORY("\TC41\data")	Selects disk in drive as AAA directory
=FILES("*.*")	Opens AAA disk
=OPEN?("*.*",0,FALSE,2)	Top of loop and Select data file from listed files
=SELECT("R38C6:R57C6")	Select data file to open
=COPY()	Copies AAA nanomole data
=CLOSE()	Closes data file
=ACTIVATE.NEXT()	
=SELECT("RC[1]")	
=PASTE()	Pastes nanomole values into worksheet
=SET.NAME("counter1",counter1+1)	Adds value of 1 to the counter1
=IF(counter1<=(!B1),GOTO(A46))	Checks to see what value = counter1 if >=1 then loops up to the top of the loop, if 0 then proceeds downward
=SELECT(!C40)	
=INPUT("What is the name of your first replicate?",2, "Name of 1st replicate","")	Asks for name of first data file chosen
=IF(A56=FALSE,HALT())	
=FORMULA(A56)	
=SELECT(!D40)	
=INPUT("What is the name of your second replicate?",2, "Name of 2nd replicate","")	Asks for name of second data file chosen
=IF(A60=FALSE,HALT())	
=FORMULA(A60)	
=SELECT(!E40)	
=INPUT("What is the name of your third replicate?",2, "Name of 3rd replicate","")	Asks for name of third data file chosen
=IF(A64=FALSE,HALT())	
=FORMULA(A64)	
=SAVE.AS?(,1)	Asks if you want to save data
=IF(A67=FALSE,HALT())	Halts macro if Cancel button is clicked
=ACTIVATE("MACRO4.XLM")	
=UNHIDE()	
=ACTIVATE.NEXT()	
=RETURN()	Halts the macro

Table 3
AAA Macro Template

A	B	C	D
AAA Macro	3		
			Protein Template
Name:		Researcher:	Protein Name:
Extension:	0		Mail Stop:
Amino Acid	Theoretical Composition	= (C66)	= (D66)
CyA	0	= C67	= D67
Asx	35	= C68	= D68
Thr	37	= C69	= D69
Ser	59	= C70	= D70
Glx	41	= C71	= D71
Pro + Cys SH	24	= C72	= D72
Gly	33	= C73	= D73
Ala	30	= C74	= D74
1/2 Cys-Cys	10	= C75	= D75
Val	35	= C76	= D76
Met	3	= C77	= D77
Ile	12	= C78	= D78
Leu	32	= C79	= D79
Nle	0	= C80	= D80
Tyr	22	= C81	= D81
Phe	13	= C82	= D82
His	7	= C83	= D83
Lys	27	= C84	= D84
Arg	12	= C86	= D86
Molecular Mass (g/mole)	47503.01		
	ul in Ampoule		
	smp recon(ul)		
	dilution factor		
column load (ul)	50		
	nMois Protein	= F62	= G62
	Protein Concentration mg/mL	= ([C33*B28*0.001*F30/B32]/F29)*F31	= ([D33*B28*0.001*F30/B32]/F29)*F31
Amino Acid	Theoretical Composition	data file #1	data file #2
CyA	= B7	0.01	0.011
Asx	= B8	5.132	5.07
Thr	= B9	5.146	5.086
Ser	= B10	7.692	7.58
Glx	= B11	5.875	5.8
Pro + CySH	= B12	4.415	4.227
Gly	= B13	4.831	4.76
Ala	= B14	4.545	4.474
1/2 Cys-Cys	= B15	1.263	1.375
Val	= B16	4.922	4.851
Met	= B17	0.39	0.365
Ile	= B18	1.68	1.662
Leu	= B19	4.68	4.812
Nle	= B20	0	0
Tyr	= B21	3.11	3.066
Phe	= B22	1.916	1.893
His	= B23	1.039	1.022
Lys	= B24	3.971	3.906
NH4	0	7.145	6.989
Arg	= B25	1.753	1.736
Total nMoles	= SUM(B42:B60)	= SUM(C41:C60)-C59	= SUM(D41:D60)-D59
Total nMoles/Total # residues		C61/B61	D61/B61

	E	F	G	H	
					1
					2
		Protein X			3
					4
	0				5
	= (E66)	Averages			6
	= E67	= AVERAGE(C7:E7)			7
	= E68	= AVERAGE(C8:E8)			8
	= E69	= AVERAGE(C9:E9)			9
	= E70	= AVERAGE(C10:E10)			10
	= E71	= AVERAGE(C11:E11)			11
	= E72	= AVERAGE(C12:E12)			12
	= E73	= AVERAGE(C13:E13)			13
	= E74	= AVERAGE(C14:E14)			14
	= E75	= AVERAGE(C15:E15)			15
	= E76	= AVERAGE(C16:E16)			16
	= E77	= AVERAGE(C17:E17)			17
	= E78	= AVERAGE(C18:E18)			18
	= E79	= AVERAGE(C19:E19)			19
	= E80	= AVERAGE(C20:E20)			20
	= E81	= AVERAGE(C21:E21)			21
	= E82	= AVERAGE(C22:E22)			22
	= E83	= AVERAGE(C23:E23)			23
	= E84	= AVERAGE(C24:E24)			24
	= E86	= AVERAGE(C25:E25)			25
					26
					27
					28
		20			29
		150			30
		1			31
					32
	= H62				33
	= ([E33*B28*0.001*F30/B32]/F29)*F31	= AVERAGE(C25:E25)			34
	data file #3	Ave nM cal	Ave nM cal	Ave nM cal	40
	0.011				41
	5.075	= C42/B42	= D42/B42	= E42/B42	42
	5.091				43
	7.586				44
	5.801	= C45/B45	= D45/B45	= E45/B45	45
	4.417				46
	4.765				47
	4.481	= C48/B48	= D48/B48	= E48/B48	48
	1.182				49
	4.86				50
	0.401				51
	1.653				52
	4.797	= C53/B53	= D53/B53	= E53/B53	53
	0				54
	3.067				55
	1.889	= C56/B56	= D56/B56	= E56/B56	56
	1.02	= C57/B57	= D57/B57	= E57/B57	57
	3.914	= C58/B58	= D58/B58	= E58/B58	58
	7.148				59
	1.75	= C60/B60	= D60/B60	= E60/B60	60
	= SUM(E41:E60)-E59				61
	= E61/B61	= AVERAGE (F42:F60)	= AVERAGE (G42:G60)	= AVERAGE (H42:H60)	62

(continued)

Table 3 (continued)

A	B	C	D
Amino Acid	Theoretical Composition	= C40	= D40
CyA	= B7	= IF(C41/[F$62]=0, " " ,[C41/F$62])	= IF(D41/[G$62]=0, " " ,[D41/G$62])
Asx	= B8	= IF(C42/[F$62]=0, " " ,[C42/F$62])	= IF(D42/[G$62]=0, " " ,[D42/G$62])
Thr	= B9	= IF(C43/[F$62]=0, " " ,[C43/F$62])	= IF(D43/[G$62]=0, " " ,[D43/G$62])
Ser	= B10	= IF(C44/[F$62]=0, " " ,[C44/F$62])	= IF(D44/[G$62]=0, " " ,[D44/G$62])
Glx	= B11	= IF(C45/[F$62]=0, " " ,[C45/F$62])	= IF(D45/[G$62]=0, " " ,[D45/G$62])
Pro + CySH	= B12	= IF(C46/[F$62]=0, " " ,[C46/F$62])	= IF(D46/[G$62]=0, " " ,[D46/G$62])
Gly	= B13	= IF(C47/[F$62]=0, " " ,[C47/F$62])	= IF(D47/[G$62]=0, " " ,[D47/G$62])
Ala	= B14	= IF(C48/[F$62]=0, " " ,[C48/F$62])	= IF(D48/[G$62]=0, " " ,[D48/G$62])
1/2 Cys-Cys	= B15	= IF(C49/[F$62]=0, " " ,[C49/F$62])	= IF(D49/[G$62]=0, " " ,[D49/G$62])
Val	= B16	= IF(C50/[F$62]=0, " " ,[C50/F$62])	= IF(D50/[G$62]=0, " " ,[D50/G$62])
Met	= B17	= IF(C51/[F$62]=0, " " ,[C51/F$62])	= IF(D51/[G$62]=0, " " ,[D51/G$62])
Ile	= B18	= IF(C52/[F$62]=0, " " ,[C52/F$62])	= IF(D52/[G$62]=0, " " ,[D52/G$62])
Leu	= B19	= IF(C53/[F$62]=0, " " ,[C53/F$62])	= IF(D53/[G$62]=0, " " ,[D53/G$62])
Nle	= B20	= IF(C54/[F$62]=0, " " ,[C54/F$62])	= IF(D54/[G$62]=0, " " ,[D54/G$62])
Tyr	= B21	= IF(C55/[F$62]=0, " " ,[C55/F$62])	= IF(D55/[G$62]=0, " " ,[D55/G$62])
Phe	= B22	= IF(C56/[F$62]=0, " " ,[C56/F$62])	= IF(D56/[G$62]=0, " " ,[D56/G$62])
His	= B23	= IF(C57/[F$62]=0, " " ,[C57/F$62])	= IF(D57/[G$62]=0, " " ,[D57/G$62])
Lys	= B24	= IF(C58/[F$62]=0, " " ,[C58/F$62])	= IF(D58/[G$62]=0, " " ,[D58/G$62])
NH4	0	= IF(C59/[F$62]=0, " " ,[C59/F$62])	= IF(D59/[G$62]=0, " " ,[D59/G$62])
Arg	= B25	= IF(C60/[F$62]=0, " " ,[C60/F$62])	= IF(D60/[G$62]=0, " " ,[D60/G$62])

2. After removal of the acid, reconstitute the samples with Li-S buffer, then analyze using lithium citrate buffers with a lithium-exchange column.
3. Use the analysis conditions given in **Table 4**. If needed, the separation between Nle and Tyr (which elutes after Nle) can be increased by lowering the column temperature.
4. Set the detector to the most sensitive scale (0.1 AUFS) (*see* **Note 17**). A chromatogram is given in **Fig. 3**.

3.4. Hydroxylysine Analysis

1. Hydrolyze samples containing 50–100 μg of protein (*see* **Note 18**) for 24 h as described in **Subheading 3.1**.
2. Remove the acid, then reconstitute the samples with Li-S buffer, and analyze using the modified program given in **Table 5** (*see* **Note 19**). The standard chromatogram is given in **Fig. 4** (*see* **Note 20**).

3.5. Carboxypeptidase Analysis

Applications involving single or combinations of carboxypeptidases to assign C-terminal protein sequences have been adequately described elsewhere (*18*).

1. Add norleucine to samples prior to the addition of carboxypeptidases at equimolar ratios (e.g., 10 nmol Nle for a sample containing 10 nmol of polypeptide).
2. Take aliquots at various time-points and place in Eppendorf tubes containing an equal volume of 2 *N* glacial acetic acid.
3. Heat for 2 min at 100°C on a boiling water bath to halt the digestion and precipitate the protein.

E	
= E40	66
= IF(E41/[H$62]=0, " " ,[E41/H$62])	67
= IF(E42/[H$62]=0, " " ,[E42/H$62])	68
= IF(E43/[H$62]=0, " " ,[E43/H$62])	69
= IF(E44/[H$62]=0, " " ,[E44/H$62])	70
= IF(E45/[H$62]=0, " " ,[E45/H$62])	71
= IF(E46/[H$62]=0, " " ,[E46/H$62])	72
= IF(E47/[H$62]=0, " " ,[E47/H$62])	73
= IF(E48/[H$62]=0, " " ,[E48/H$62])	74
= IF(E49/[H$62]=0, " " ,[E49/H$62])	75
= IF(E50/[H$62]=0, " " ,[E50/H$62])	76
= IF(E51/[H$62]=0, " " ,[E51/H$62])	77
= IF(E52/[H$62]=0, " " ,[E52/H$62])	78
= IF(E53/[H$62]=0, " " ,[E53/H$62])	79
= IF(E54/[H$62]=0, " " ,[E54/H$62])	80
= IF(E55/[H$62]=0, " " ,[E55/H$62])	81
= IF(E56/[H$62]=0, " " ,[E56/H$62])	82
= IF(E57/[H$62]=0, " " ,[E57/H$62])	83
= IF(E58/[H$62]=0, " " ,[E58/H$62])	84
= IF(E59/[H$62]=0, " " ,[E59/H$62])	85
= IF(E60/[H$62]=0, " " ,[E60/H$62])	86

4. After cooling on wet ice, centrifuge the samples and transfer the supernatant to another Eppendorf tube.
5. Dry by rotary evaporation using a Savant SpeedVac.
6. Reconstitute the samples with Li-S buffer.
7. Follow the operating conditions given in **Table 6**. The initial 40-min segment of the chromatogram obtained using this modified program is shown in **Fig. 5**.

3.6. Amino Sugar Analysis (see Note 21)

1. Divide samples containing 2–20 µg of protein or 0.5–5 nmol of peptide fractions into two identical aliquots.
2. Hydrolyze one aliquot for 24 h at 110°C as described in **Subheading 3.1.**
3. Hydrolyze the other aliquot for only 2 h at 110°C (*see* **Note 22**).
4. After removal of the acid, reconstitute the 2-h hydrolysates with Na-S buffer and analyze using a modified program given in **Table 7** (*see* **Note 23**).
5. Analyze the 24-h hydrolysates using the standard method (**Table 1**) (**Fig. 6**) for quantitation of the protein/peptide to permit molar GlcNAc and/or GalNAc determinations. Tryptophan standards should also be analyzed to ensure that Trp does not coelute with GalNH$_2$; if necessary, this resolution can be improved by lowering the column temperature (*see* **Note 24**).

3.7. Cysteine Alkylation Monitoring

1. Hydrolyze desalted samples containing 2–20 µg of S-carboxy-methylated or S-carboxyamidomethylated proteins for 24 h as described in **Subheading 3.1.**
2. After removal of the acid, reconstitute the samples with Na-S buffer, and analyze using the standard program (**Table 1**) (*see* **Note 25**). Representative chromato-

AAA Macro	3					
		Protein Template				
Name:	Researcher		Protein Name	Protein X		
Extension:	0		Mail Stop:	0		
Amino Acid	Theoretical Composition	data file #1	data file #2	data file #3	Averages	
CyA	0	0.1	0.1	0.1	0.1	
Asx	35	34.7	34.8	34.8	34.7	
Thr	37	34.8	34.9	34.9	34.9	
Ser	59	52.0	52.0	52.0	52.0	
Glx	41	39.7	39.8	39.7	39.7	
Pro + Cys SH	24	29.9	29.0	30.3	29.7	
Gly	33	32.7	32.6	32.6	32.7	
Ala	30	30.7	30.7	30.7	30.7	
1/2 Cys-Cys	10	8.7	9.4	8.1	8.7	
Val	35	33.3	33.3	33.3	33.3	
Met	3	2.6	2.5	2.7	2.6	
Ile	12	11.4	11.4	11.3	11.4	
Leu	32	33.0	33.0	32.9	33.0	
Nle	0					
Tyr	22	21.0	21.0	21.0	21.0	
Phe	13	13.0	13.0	12.9	13.0	
His	7	7.0	7.0	7.0	7.0	
Lys	27	26.9	26.8	26.8	26.8	
Arg	12	11.9	11.9	12.0	11.9	
Molecular Mass (g/mole)	47503.01					
	ul in Ampoule				20	
	smp recon(ul)				150	
	dilution factor				1	
column load (ul)	50					
	nMols Protein	0.148	0.146	0.146		
	Protein Concentration mg/mL	1.054	1.039	1.040	1.044	

Fig. 2. Summary sheet using the AAA macro. Average compositions are given in the right-hand column, and the average mg/mL value is provided in the lower right hand box. CyA refers to cysteic acid, which is present when samples are oxidized intentionally.

grams for the standard mixture containing carboxymethylcysteine (CMCys) and for an *S*-carboxymethylated recombinant antibody sample are given in **Fig. 7A,B**, respectively.

3. Monitor CMCys and half-cystine residue/mol values to determine the extent of cysteine alkylation (*see* **Notes 26** and **27**).

Table 4
Norleucine Analysis

Time (min)	Event	Conditions
0.0	Sample injection	Li-A buffer, 38°C
44.0	Buffer change	Li-A to Li-D
73.0	Buffer change	Li-D to Li-R
74.0	Reagent pump	Ninhydrin to water
77.0	Buffer change	Li-R to Li-A
80.0	Reagent pump	Water to ninhydrin
90.0	Recyle (start next run)	

Buffer pump: 20 mL/h.
Reagent pump: 10 mL/h.

4. Notes

1. Precolumn derivatization with RP-HPLC separation is used for amino acid analysis; a popular version is the Waters AccQTag system *(21)*. These precolumn methods may not be suitable for detection of trace levels of minor amino acids (the needle-in-a-haystack problem) because the peak resolutions are diminished when the sample loads are increased, whereas resolution is maintained with higher loads using the cation–exchange systems. In addition, precolumn accuracy may be limited if derivatization is incomplete, a problem that does not occur with postcolumn derivatization systems. Similarly, cation–exchange systems are more tolerant of salts and residual HCl than the precolumn systems.

2. Hydrolysis ampoules are wrapped in heavy duty foil and pyrolyzed by heating for 24 h at 400°C in a muffle furnace before use.

3. Production of the Beckman 6300 analyzers described in this chapter has been halted, but a similar system can be fashioned using components offered by Pickering Labs (Mountain View, CA) *(19)*. Pickering also supplies amino acid analysis buffers, reagents, and columns for the Beckman 6300, but care must be taken not to combine Pickering's Trione ninhydrin reagent with mobile phases that contain alcohols (such as Beckman's Na-A and Na-B) as this combination may clog the analyzer's reactor.

4. Hitachi also offers a cation–exchange amino acid analysis instrument.

5. Dionex has recently introduced an anion–exchange system that detects underivatized amino acids using pulsed amperometric detection *(20)* (*see also* Chapter 7, this volume), but we have no experience with this system.

6. A wash bottle containing 1 *M* sodium bicarbonate is kept nearby wherever HCl ampoules are opened to neutralize spills.

7. The prepared standards should be stored refrigerated in aliquots using screw-top Eppendorf tubes equipped with a rubber gasket to prevent evaporation. Tryptophan tends to degrade over time in acid conditions, so fresh Trp standards should be prepared when needed; commercial preparations containing Trp may not be reliable.

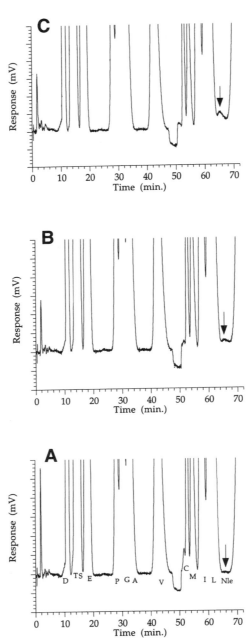

Fig. 3. Analysis for norleucine incorporation at Met positions. Aliquots from 40 μg of a recombinant protein are given, with the arrow indicating the Nle peak after additions of **(A)** 0 pmol Nle **(B)** 200 pmol Nle, or **(C)** 400 pmol Nle. Operating parameters are given in **Table 4**.

Table 5
Hydroxylysine Analysis

Time (min)	Event	Conditions
0.0	Sample injection	Li-A buffer, 38°C
12.0	Temperature change	38°C to 50°C over 8 min
42.0	Buffer change	Li-A to Li-B
60.0	Temperature change	50°C to 71°C over 8 min
70.0	Buffer change	Li-B to Li-C
110.0	Reagent pump	Ninhydrin to water
111.0	Buffer change	Li-C to Li-R
114.0	Buffer change	Li-R to Li-A
125.0	Reagent pump	water to ninhydrin
135.0	Recyle (start next run)	

Buffer pump: 20 mL/h.
Reagent pump: 10 mL/h.

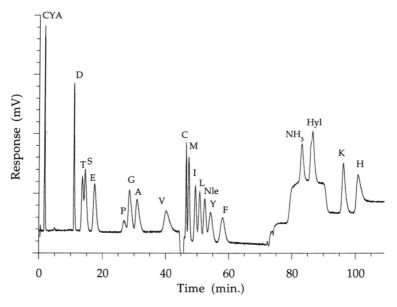

Fig. 4. Analysis for hydroxylysine. A standard mixture containing 1 nmol of each component was loaded. Hyl appears as a poorly-resolved peak pair. Operating parameters are given in **Table 5**.

8. Protein/peptide quantitation can be compromised by multiple sample transfers. When accuracy is essential, samples should be transferred directly from the primary container to the hydrolysis ampoule.

Table 6
Analysis of Carboxypeptidase Supernatants

Time (min)	Event	Conditions
0.0	Sample injection	Li-A buffer, 38°C
12.0	Temperature change	38°C to 50°C over 8 min
43.0	Buffer change	Li-A to Li-B
60.0	Temperature change	50°C to 73°C over 8 min
60.0	Buffer change	Li-B to Li-C
130.0	Reagent pump	Ninhydrin to water
132.0	Buffer change	Li-C to Li-R
134.0	Buffer change	Li-R to Li-A
140.0	Temperature change	73°C to 38°C
140.0	Reagent pump	Water to ninhydrin
155.0	Recyle (start next run)	

Buffer pump: 20 mL/h.
Reagent pump: 10 mL/h.

Table 7
Amino Sugar Analysis

Time (min)	Event	Conditions
0.0	Sample injection	Na-F buffer, 66°C
55.0	Reagent pump	Ninhydrin to water
57.0	Buffer change	Na-F to Na-R
58.5	Buffer change	Na-R to Na-F
60.0	Reagent pump	Water to ninhydrin
74.0	Recyle (start next run)	

Buffer pump: 20 mL/h.
Reagent pump: 10 mL/h.

9. Alternative hydrolysis systems have been proposed, including the Waters PicoTag batch hydrolysis system, in which the 6 *N* HCl is placed outside the sample tubes in a chamber that can be evacuated, closed, and heated. Phenol must also be added to prevent destruction of tyrosine. This system has the advantage that direct contact with the acid is avoided, eliminating a potential source of contamination, but in our experience the poor hydrolysis of Ile-Ile, Ile-Val, and Val-Val bonds with vapor–phase hydrolysis makes this technique unsuitable.

10. Hydrolysis at 155°C for 60 min has also been proposed, but we seldom use this procedure because Thr and Ser values are greatly reduced. Also, because we typically batch samples together, a 24-h hydrolysis is often more convenient from an operational standpoint.

11. A Perkin Elmer Model 200 autosampler has replaced the original coil system, with a fixed 50-μL volume used for standards and samples.

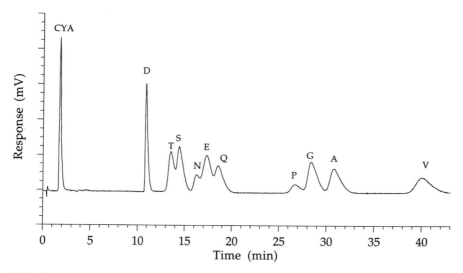

Fig. 5. Analysis of carboxypeptidase digestion samples (expanded view). A standard mixture containing 1 nmol of each component was loaded. For clarity, only the early region of the chromatogram is provided to show the elution positions of Asn and Gln; the complete chromatogram is essentially the same as **Fig. 8**. Operating parameters are given in **Table 6**.

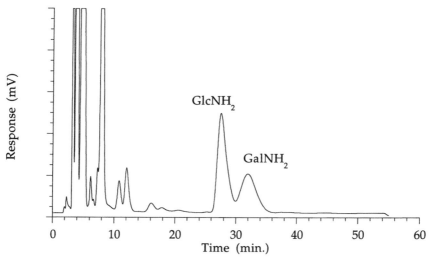

Fig. 6. Amino sugar analysis. A hydrolysate from 1 nmol of a recombinant glycoprotein was loaded. Operating parameters are given in **Table 7**.

12. In the tables, the "reagent pump" event refers to changing the solution added postcolumn from a ninhydrin-containing reagent to water (or vice versa); this is done to avoid having NaOH (Na-R) or LiOH (Li-R) mix with the ninhydrin reagent.

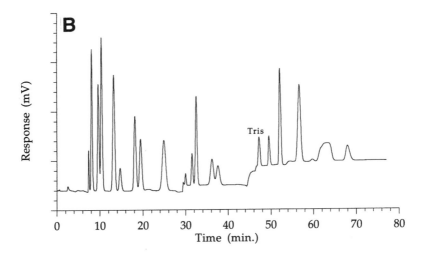

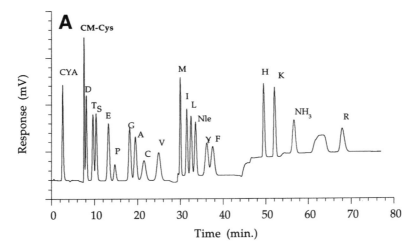

Fig. 7. Carboxymethylcysteine analysis. (**A**) Standard mixture containing 2 nmol of each component. (**B**) Analysis of 5 μg of a recombinant antibody after reduction and S-carboxymethylation. Operating parameters are given in **Table 1**. The peak eluting at approx 47 min is Tris buffer.

13. Key amino acids that typically provide quantitative recoveries (e.g., Asx, Glx, Ala, Leu, Phe, His, Lys, Arg) are used to determine the total nmol of protein or peptide (in the example provided for data file #1 in **Table 3**, 5.132 nmol of Asx are divided by 35 residues of Asx expected per mol of protein to produce a nmol

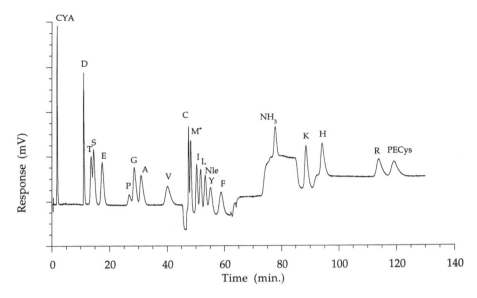

Fig. 8. Pyridylethylcysteine analysis. Analysis of a standard mixture containing 1 nmol of each component. Operating parameters are given in **Table 6**.

protein value); use of these selected amino acids avoids the low recoveries experienced for the acid-labile amino acids, especially Trp, Cys, Thr, and Ser *(1)*.

14. The amino acid nmol values are divided by the total protein/peptide nmol value to produce residues/mol values that are averaged in the right-hand column (in the example provided in **Table 3** and **Fig. 2**, the nmol values in rows 42–60 of columns C, D, and E are divided by the average protein value of 0.148, 0.146, and 0.146, respectively, then averaged). The nmol protein/peptide values in **Fig. 2** are also multiplied by the molecular mass to provide the total µg injected on the analyzer, which is then converted to a µg/µL value (same as a mg/mL value) by dividing by the volume loaded in the ampoule (e.g., 20 µL in **Fig. 2**), then correcting for the reconstitution volume and µL injected (e.g., multiplied by 150/50 in **Fig. 2**); these concentrations are then averaged as shown in the lower right-hand box of **Fig. 2**.

15. When the protein/peptide composition is not known, or if the sample contains a mixture of proteins, then quantitation is performed by summing the nanograms contributed by each amino acid residue. For example, instead of calibrating to a 2 nmol standard value for alanine, the data system calibrates to 142 ng (equivalent to 2 nmol of alanine using the residue mass). The values for all amino acids are summed. This method generates values that are usually 5–10% lower than their true quantity because of poor recoveries of acid-labile amino acids, but in our experience these values are likely to be more sensitive and reliable than colorimetric methods.

16. Norleucine quantitation is difficult at levels below 1% Nle-for-Met replacement. Nle added at several levels to samples can establish the lowest level of quantitation, which for us is typically about 75 pmol. In our experience, the incorporation of Nle occurs proportionately at every Met position; therefore, the mol Nle per mol protein value can be divided by the number of methionines to provide percent Nle-for-Met replacement values. Nle replacement can sometimes be observed by electrospray mass spectrometry of intact proteins (18 Dalton lower mass), provided that no other sources of heterogeneity are present (V. Ling, unpublished data).

17. Most amino acids will be present in great abundance, saturating the detector, but the glycine and proline peaks are usually still on-scale, and thus can be used for mol Nle per mol protein quantitation. Chromatograms for 40 μg hydrolysates of a recombinant protein spiked with 0, 200, or 400 pmol Nle are given in **Fig. 3A**, **B**, and **C**, respectively.

18. Sensitive detection of hydroxylysine (Hyl) is difficult because of the fact that most noncollagenous molecules are at most partially modified at just one -Lys-Gly- positio n, thus the overall percentage of modified Lys is very low.

19. In the standard amino acid analysis method (**Table 1**), Hyl coelutes with histidine, so it would not be observed in an intact protein that contains His, and it might be misinterpreted in a peptide fraction. The long delay for the second buffer change is needed to increase the resolution of Hyl from ammonia. Hyl appears as a partially resolved doublet peak due to racemization of the δ carbon during hydrolysis; therefore, the peak areas are summed.

20. Once it has been determined that Hyl is present, peptide maps may be used to assign the site provided the investigator is aware that Hyl–Gly bonds are fairly resistant to trypsin and endoproteinase Lys-C digestion *(22)*.

21. Amino sugar analysis does not provide complete monosaccharide determinations of the types obtained by techniques such as HPAEC-PAD or GC-MS, but it does have the advantages that no additional equipment is required, and the results are routinely quantitative. This approach is most useful when assaying proteins for the mucin-type O-linked oligosaccharides that contain GalNAc at their reducing termini. O-linked structures that lack GalNAc are rare, but have been found in EGF-like domains of several glycoproteins *(23)*. Some N-linked structures contain GalNAc, particularly in proteins from human embryonic kidney (293) cells *(24)* or melanoma cells *(25)*, but these N-linked structures can be released using PNGaseF to allow discrimination between N-linked and O-linked GalNAc residues.

22. After 2 h of hydrolysis, GlcNAc is hydrolyzed to $GlcNH_2$, whereas GalNAc is hydrolyzed to $GalNH_2$, and both are released quantitatively.

23. This program starts with the second buffer used in the standard analysis, so most amino acids elute near the beginning of the chromatogram. A chromatogram from 1 nmol of a hydrolysate of a recombinant glycoprotein containing both N-linked and O-linked sites is given in **Fig. 6**.

24. Proteins that are highly glycosylated will have some residual amino sugars that will appear as a broad peak that elutes in the Ile-Leu-Nle region of the standard chromatogram. Increasing the hydrolysis time to 72 h will eliminate this peak.
25. Methionine residues can also be unintentionally S-alkylated, but this can be detected by the presence of trace levels of homoserine, a hydrolysis product of S-carboxymethylmethionine that elutes between Ser and Glx.
26. Samples that have been alkylated using 4-vinylpyridine need to be analyzed using the lithium citrate program that is used for the carboxypeptidase digestion samples (**Table 6**). Pyridylethylcysteine (PECys) is very basic, and elutes after Arg (**Fig. 8**).
27. When monitoring Cys alkylation conditions, attention should be paid to methionine recoveries, as the conditions (such as trace metals or residual O_2) that affect Met recoveries will also affect CMCys recoveries. In addition, methionine sulfoxide can coelute with CMCys; therefore, samples should be analyzed promptly after acid removal.

References

1. Moore, S. and Stein, W. H. (1958) Chromatographic determination of amino acids by the use of automatic recording equipment. *Methods Enzymol.* **6,** 819–831.
2. Schuster, R. (1988) Determination of amino acids in biological, pharmaceutical, plant and food samples by automated precolumn derivatization and high performance liquid chromatography. *J. Chromatog.* **431,** 217–284.
3. Heinrickson, R. L. and Meredith, S. C. (1983) Amino acid analysis by reverse-phase high-performance liquid chromatography: precolumn derivatization with phenylisothiocyanate. *Anal. Biochem.* **136,** 65–74.
4. van Wandlen, C. and Cohen, S. A. (1997) Using quaternary high-performance liquid chromatography eluent systems for separating 6-aminoquinolyl-N-hydroxysuccinimidyl carbamate-derivatized amino acid mixtures. *J. Chromatog. A* **763,** 11–22.
5. Kisumi, M., Sugiura, M., and Chibata, I. (1976) Biosynthesis of norvaline, norleucine and homoisoleucine in Serratia marcescens. *J. Biochem.* **80,** 333–339.
6. Tsai, L. B., Lu, H. S., Kenney, W. C., Curless, C. C., Klein, M. L., Lai, P.-H., et al. (1988) Control of misincorporation of de novo synthesized norleucine into recombinant interleukin–2 in E. coli. *Biochem. Biophys. Res. Commun.* **156,** 733–739.
7. Bogosian, G., Violand, B. N., Dorward-King, E. J., Workman, W. E., Jung, P. E., and Kane, J. F. (1989) Biosynthesis and incorporation into protein of norleucine by Escherichia coli. *J. Biol. Chem.* **264,** 531–539.
8. Kivirikko, K. I., Myllyla, R., and Pihlajaniemi, T. (1992) Hydroxylation of proline and lysine residues in collagens and other animal and plant proteins, in *Posttranslational Modifications of Proteins* (Harding, J. J. and Crabbe, M. J., eds.), CRC, Boca Raton, FL, pp. 1–51.
9. Molony, M. S., Wu, S.-L., Keyt, L., and Harris, R. J. (1995) The unexpected presence of hydroxylysine in non-collagenous proteins, in *Techniques in Protein Chemistry VI* (Crabbe, J., ed.), Academic, San Diego, CA, pp. 91–98.

10. Kornfeld, R. and Kornfeld, S. (1985) Assembly of asparagine-linked oligosaccharides. *Annu. Rev. Biochem.* **54,** 631–664.
11. Tarentino, A. L., Gomez, C. M., and Plummer, T. H. (1985) Deglycosylation of asparagine-linked glycans by peptide: N-glycosidase F. *Biochemistry* **24,** 4665–4671.
12. O'Connell, B., Tabak, L. A., and Ramasubbu, N. (1991) The influence of flanking sequences on O-glycosylation. *Biochem. Biophys. Res. Commun.* **180,** 1024–1030.
13. Wilson, I. B. H., Gavel, Y., and von Heijne, G. (1991) Amino acid distributions around O-linked glycosylation sites. *Biochem. J.* **275,** 528–534.
14. Pisano, A., Packer, N. H., Redmond, J. W., Williams, K. L., and Gooley, A. A. (1994) Characterization of O-linked glycosylation motifs in the glycopeptide domain of bovine κ-casein. *Glycobiology* **4,** 837–844.
15. Garnick, R. L., Solli, N. J., and Papa, P. A. (1988) The role of quality control in biotechnology: an analytical perspective. *Anal. Chem.* **60,** 2546–2557.
16. Lundell, N. and Schreitmüller, T. (1999) Sample preparation for peptide mapping — a pharmaceutical quality-control perspective. *Anal. Biochem.* **266,** 31–47.
17. Jones, M. D., Merewether, L. A., Clogston, C. L., and Lu, H. S. (1994) Peptide map analysis of recombinant human granulocyte stimulating factor: elimination of methionine modification and nonspecific cleavages. *Anal. Biochem.* **216,** 135–146.
18. Allen, G. (1989) Determination of the carboxy-terminal residue, in *Sequencing of Proteins and Peptides*, Elsevier, Amsterdam and New York, pp. 67–71.
19. Grunau, J. A. and Swaider, J. M. (1992) Chromatography of 99 amino acids and other ninhydrin-reactive compounds in the Pickering lithium gradient system. *J. Chromatog.* **594,** 165–171.
20. Clarke, A. P., Jandik, P., Rocklin, R. D., Liu, Y., and Avdalovic, N. (1999) An integrated amperometry waveform for the direct, sensitive detection of amino acids and amino sugars following anion-exchange chromatography. *Anal. Chem.* **71,** 2774–2781.
21. Strydom, D. J. (1996) Amino acid analysis using various carbamate reagents for precolumn derivatization, in *Techniques in Protein Chemistry VII* (Marshak, D. R., ed.), Academic, San Diego, CA, pp. 331–339.
22. Molony, M. S., Quan, C., Mulkerrin, M. G., and Harris, R. J. (1998) Hydroxylation of Lys residues reduces their susceptibility to digestion by trypsin and lysyl endopeptidase. *Anal. Biochem.* **258,** 136–137.
23. Harris, R. J. and Spellman, M. W. (1993) O-Linked fucose and other post-translational modifications unique to EGF modules. *Glycobiol.* **3,** 219–224
24. Yan, S. B., Chao, Y. B., and van Halbeek, H. (1993) Novel Asn-linked oligosaccharides terminating in GalNAcβ(1→4)[Fucα(1→3)]GlcNAcβ(1→•) are present in recombinant human Protein C expressed in human kidney 293 cells. *Glycobiology* **3,** 597–608.
25. Chan, A. L., Morris, H. R., Panico, M., Eteinne, A. T., Rogers, M. E., Gaffney, P., et al. (1991) A novel sialylated N-acetylgalactosamine-containing oligosaccharide is the major complex-type structure present in Bowes melanoma tissue plasminogen activator. *Glycobiology* **1,** 173–185.

3

Purification of Proteins Using UltraMacro Spin Columns or ProSorb Sample Preparation Cartridges for Amino Acid Analysis

Li Zhang and Nancy Denslow

1. Introduction

Amino acid analysis of proteins is one of the best and most accurate methodologies to quantify proteins. Ideally, samples should be pure prior to hydrolysis, however, they are often not only at low concentration, but also in buffers, salts, and/or detergents. Sample contaminants contribute to background noise that hampers amino acid analysis, variably affecting recoveries of individual amino acids *(1)*. The necessity to obtain samples in high concentration and free of contaminants is a problem for protein quantitation and structural characterization (*see* **Note 1**). Many of the existing methods for removing buffers and salts from proteins, such as precipitation methods or reverse-phase high-performance liquid chromatography (RP-HPLC) result in variable and unacceptably high losses of protein (*see* **Note 2**), making an assessment of the true concentration difficult. In this chapter, we describe two methods commonly used to recover proteins from buffers: a gel filtration method, Ultra microspin columns, and adsorption to a polyvinylidene difluoride (PVDF) membrane in the ProSorb sample preparation cartridge from Applied Biosystems. Both methods work well, with somewhat better recoveries for the gel filtration method.

2. Materials
2.1. Equipment

1. UltraMicro spin (gel filtration) columns (Amika Corp, Columbia, MD) are available in two sizes (*see* **Note 3**):

From: *Methods in Molecular Biology*, vol. 159: *Amino Acid Analysis Protocols*
Edited by: C. Cooper, N. Packer, and K. Williams © Humana Press Inc., Totowa, NJ

	Sample vol	Bed vol
Ultramicrospin column	5–25 mL	100 µL
Macrospin column	50–150 µL	500 µL

2. ProSorb sample preparation cartridge (Applied Biosystems, Foster City, CA).
3. Immobilon-PSQ transfer membrane (Millipore Corp., Bedford, MA).
4. Sample hydrolysis tubes (Corning Inc., Corning, NY) (*see* **Note 4**).
5. Reaction Vials (Waters Associates, Milford, MA).

2.2. Reagents

1. HPLC-grade water (*see* **Note 5**).
2. Trifluoroacetic acid (TFA).
3. Methanol.
4. Hydrochloric acid (HCl).
5. Phenol.

2.3. Reagents

1. 20% methanol, 0.1 N HCl.
2. 6 N HCl + 0.1% phenol.
3. 0.1% TFA.

3. Methods

3.1. Preparation of the Macrospin column (sample 50–150 µL)

1. The macrospin columns should be prepared basically as suggested by the manufacturer (*2*), except that we have introduced several more washes to better equilibrate the columns. In brief, tap the column gently to recover the gel material at
   the bottom of the column. Hydrate the gel with 500 µL of HPLC-grade water for
   15 min at room temperature (*see* **Note 6**) and then centrifuge the column for
   4 min at about 2000g to equilibrate the column.
2. Wash the column 3 times with 500 µL of water, to ensure full equilibration.
3. Blot the exterior of the column dry to remove all moisture and place it in a new
   collecting tube. For the macrospin columns, add 50–150 µL of sample to the top,
   being careful to place the sample in the center of the column.
4. Centrifuge the tube for 4 min at 2000g, and wash the column with at least 200 µL
   of water (*see* **Note 7**).
5. Remove the purified sample and speed-vacuum to dryness before hydrolysis (*see*
   Note 8).

3.2. Ultra Microspin Columns (sample 5–25 µL)

1. As for the macrocolumns, tap the dry gel to the bottom of the column. Wet with
   50 µL of water and centrifuge it for 3 min, at 1000g to equilibrate the column. For
   best results, wash the column three times with 200 µL of water. To centrifuge
   these columns, place them directly into a microcentrifuge or use an empty 1.7-
   mL microfuge tube as an adapter.

Table 1
Maximum Detergent Concentrations[a]

Detergent	Maximum concentration (%) without interference to binding ProSorb PVDF membranes (v/v or w/v)
Triton X-100 (reduced)	0.01
Tween-20	0.01
SDS	0.02
Octyl glucoside	0.25
Brij 35	0.02

[a]Data from PE Biosystems (1996) *(7)*.

2. After the column has been equilibrated, blot dry to remove any moisture that may be adhering to the exterior of the column.
3. Add the 5–25 µL sample to the top of the spin column, being careful to place the sample in the center of the column.
4. Place the column in a new collecting tube, spin the tube for 3 min at 1000*g*. After centrifugation, the purified sample will be in the collecting tube and will be ready for further use.

3.3. Preparation of the ProSorb Sample Cartridge

1. Make sure sample is in 0.1% TFA for optimum results (*see* **Note 9**). If the sample volume is less than 100 µL, dilute sample to 100 µL with 0.1% TFA. If the sample has detergent, dilute such that the detergent concentration is below the values shown in **Table 1** (*see* **Notes 10** and **11**).
2. Place ProSorb device in an appropriate test tube rack.
3. Wet the PVDF membrane in the reservoir with 10 µL of methanol (*see* **Note 12**).
4. Add sample to reservoir (100 to 400 µL volume), and allow the sample to pass through the membrane (*see* **Note 13** and **14**).
5. Ensure that the sample reservoir has intimate contact with the absorbent and fluid transfer begins.
6. Add additional sample aliquots, if necessary (change absorbent after each 750 µL).
7. Wash the membrane with 100 µL of 0.1% TFA two times.
8. Remove sample reservoir and air-dry the PVDF membrane.
9. Punch out PVDF membrane, then continue with the next processing steps.

3.4. Hydrolysis of Sample in Solution

There are several methods that can be used to hydrolyze the sample including vapor phase hydrolysis *(3)*, hydrolysis in the tube *(4)*, and microwave hydrolysis *(5)*. All work about equally well. A vapor phase method that works very well is described below.

1. Place 6×50 mm sample tubes in a bulge modified 40-mL screw-cap vial containing 200 µL of 6 *N* HCl plus 1% phenol. Seal the vial with a modified minaret slide valve.
2. Evacuate briefly (20 s) then flush with nitrogen for 5 s. Repeat this alternating process three times. Close the slide valve on the fourth vacuum step. Heat 24 h at 120°C.
3. Add sample dilution buffer and transfer the dilution to a spin filter tube, centrifuge at 2000*g* for 4 min and store at 4°C until ready for AAA.

3.5. Hydrolysis for PVDF Membrane

There have been several methods described to improve the recovery of a sample hydrolyzed on a PVDF membrane *(4)* (*see* **Note 15**). We routinely use Immobilon-PSQ transfer membrane (Millipore Corp.).

1. Push the PVDF membrane into the bottom of a hydrolysis tube.
2. Hydrolyze as normal. After the acid has been dried from the hydrolyzed sample, add 10 µg of methanol. Extract two times with 50 µL of 20% MeOH, 0.1 *N* HCl, and transfer to a labeled 0.5-mL Eppendorf tube.
3. Put the Eppendorf tube into a dry PICOTAG vessel and dry it down.
4. Dilute with sample dilution buffer and filter using a spin column. Sample is ready for AAA.

3.6. Amino Acid Analysis

Amino acid analysis can be performed either by precolumn derivatization with phenylisothiocyanate on the Applied Biosystem 420 or by postcolumn derivation with Ninhydrin after ion–exchange separation on the Beckman 6300 high performance analyzer.

4. Notes

1. A common source of amino acid contamination is fingerprints. Be sure to wear gloves to keep this type of contamination away from the samples. Another source of contamination is dust.
2. Every time a sample is handled, a portion is lost. Minimize sample handling as much as possible.
3. It is important to choose the appropriate size of microspin column for your sample volume. For sample volumes of 5–25 µL, the smaller ultramicrospin columns are most appropriate. They have a packed bed volume of about 100 µL. Larger volumes, 50–150 µL, should be applied to the larger macrospin columns, which have a packed-bed volume of 500 µL.
4. All glassware used for hydrolysis should be thoroughly cleaned and pyrolized at 500°C.
5. Always use the highest quality of reagents (analytical grade) and water. We recommend MilliQ water.

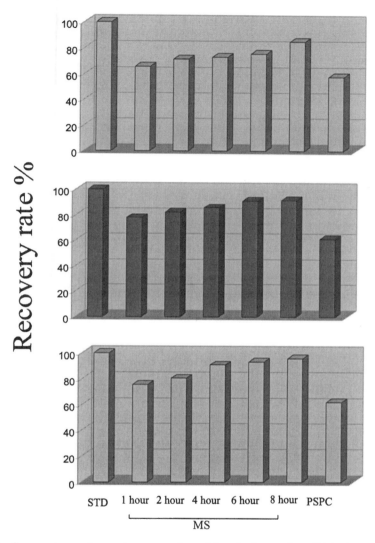

Fig. 1. Comparison of protein recoveries of desalted samples. CAII, TIM, or BSA were prepared in either H2O (std) or in a buffer containing 50 mM Tris, pH 8.0, and 0.2% SDS and desalted through either a Macrospin Column (MSC) or a ProSorb Sample Preparation Cartridge (PSPC) as indicated. Samples were hydrolyzed and analyzed on a Beckman 6300 Amino Acid Analyzer. Each bar represents average recovery from three replicates.

6. The percentage of sample recovery increases with increased hydration time for the spin column. When columns are hydrated for 4 h or longer, it is feasible to recover more than 90% of the sample (**Fig. 1**).

7. It is wise to monitor the recovery of the protein by absorbance of the column eluent at 280 nm. Collect sample until all of the material is eluted.

8. When samples containing a high concentration of protein or peptide are available in reasonable volume, there is no need to purify the sample further, simply dilute the sample with HPLC-grade water to a salt concentration that will not interfere with the analysis.

9. Acetonitrile (ACN) concentrations greater than 15% interfere with sample binding to PVDF. We recommend, therefore, that HPLC fractions be diluted with distilled water or 0.1% TFA to reduce the concentration of ACN to less than 15%, prior to using the ProSorb cartidge.

10. Often proteins submitted for amino acid analysis contain detergents used in their preparation. Detergents can dramatically affect the binding of proteins to PVDF membranes. **Table 1** summarizes the concentrations of detergents that are tolerable in a sample prior to binding to the membrane. Samples containing higher than these concentrations should be diluted to reduce the detergent and then added to the membrane in batches.

11. Many samples contain significantly more detergent than the amounts listed in **Table 1**. One way to deal with higher amounts of detergent is to add a small amount of methanol to the sample before loading it into the ProSorb cartridge. This will weaken sample interaction with the detergent.

12. ProSorb cartridges should be prepared as directed by the manufacturer, using MilliQ water for all wash steps to reduce persistent background levels of contaminating amino acids.

13. Be careful with the manipulation of samples at every step. Be aware that proteins and peptides can irreversibly bind to glass and/or plastic tubes. If your sample is small, you can minimize losses in the cartridge by adding 100 µL of buffer to the prewetted membrane prior to the addition of sample. The sample can be added directly to this buffer.

14. Samples or washes larger than 400 µL are accommodated by loading multiple aliquots into the reservoir. The PVDF membrane should not be allowed to dry between aliquots.

15. Based on a recent ABRF study on amino acid analysis *(8)*, samples bound to PVDF membranes tend to have higher error rates (20%) than those analyzed free in solution (12%). These error rates are averages for the study.

References

1. Dupont, D. R., Keim, P. S., Chui, A. H., Bello, R., Bozzini, M., and Wilson, K. J. (1989) A comprehensive approach to amino acid analysis, in *Techniques in Protein Chemistry* (Hugli, T. E., ed.). Academic, London and New York, pp. 284–294.

2. Instructions for UltraMicro spin column, Amika Corp.

3. West, K. A. and Crabb, J. W. (1992) Application of automatic PTC amino acid analysis, in *Techniques in Protein Chemistry III* (Angeletti, R. H., ed.), Academic, London and New York, pp. 233–242.

4. Gharahdaghi, F., Atherton, D., DeMott, M., Mische, S. M. (1992) Amino acid analyis of PVDF bound proteins, in *Techniques in Protein Chemistry III* (Angeletti, R. H., ed.), Academic, London and New York, pp. 249–260.
5. Gilman, L. B. and Woodward, C. (1990) Hydrolysis methods in current research, in *Protein Chemistry* (Villafranca, J. J., ed.), Academic, London and New York, pp. 23–26.
6. User bulletin No. 64, ProSorb sample preparation cartridge, January (1996).
7. Crabb, J. W., West, K. A., Dodson, W. S., and Hulmes, J. D. (1997) *Amino Acid Analysis in Current Protocols in Protein Science* (Coligan, J. E., Dunn, B. M., Ploegh, H. L., Speicher, D. W., and Wingfield, P. T., eds.), Wiley, New York, pp. 11.9.1–11.9.42.
8. Hunziker, P., Anderson, T. T., Bao, Y., Cohen, S. A., Denslow, N. D., Hulmes, J. D., et al. (1999) Identification of proteins electroblotted to polyvinylidene difluoride membrane by combined amino acid analysis and bioinformatics: an ABRF Multicenter Study. *J. Biomolecular Techniques* **10,** 129–136.

4

Amino Acid Analysis Using Precolumn Derivatization with 6-Aminoquinolyl-*N*-Hydroxysuccinimidyl Carbamate

Steven A. Cohen

1. Introduction

Over the past 20 years, amino acid analysis methods based on precolumn derivatization procedures have become a popular alternative to the traditional postcolumn derivatization techniques based on ion–exchange separation of the underivatized amino acids. The major reasons for this include greater sensitivity, faster analyses, and the ability to use less expensive, more flexible HPLC instrumentation rather than dedicated amino acid analyzers.

Among the more widely employed derivatization reagents are orthophthalaldehyde (OPA) *(1)* and Edman's reagent, phenylisothiocyanate (PITC) *(2–5)*, for which there are numerous publications illustrating their utility. OPA, however, has the significant drawback in that it does not react with secondary amino acids, and some of the derivatives, notably those with Gly and Lys, are unstable. PITC, although reactive with the secondary amino acids, also produces somewhat unstable derivatives, notably with Asp and Glu, which slowly cyclize from the desired phenylthiocarbamyl moieties to their respective phenylthiohydantion derivatives. In addition, excess PITC must be removed, usually by evaporation, prior to HPLC analysis to avoid column contamination and poor chromatographic separation.

More recently, Cohen and Michaud introduced 6-aminoquinolyl-*N*-hydroxysuccinimidyl carbamate (AQC) for precolumn derivatization *(6)*. This reagent reacts smoothly with primary and secondary amines to produce stable unsymmetrical urea derivatives that are highly fluorescent. Products are generated within seconds, and in a somewhat slower reaction, excess reagent is

From: *Methods in Molecular Biology*, vol. 159: *Amino Acid Analysis Protocols*
Edited by: C. Cooper, N. Packer, and K. Williams © Humana Press Inc., Totowa, NJ

hydrolyzed to yield 6-aminoquinoline (AMQ), carbon dioxide, and *N*-hydroxysuccinimide. Key to the design and implementation of this method are the fluorescence properties of the derivatized amino acids and AMQ. Although both have excitation maxima approx 248 nm, they have radically different emission maxima, with the AMQ near 520 nm, and the amino acid products at approx 395 nm. This shift in emission maximum allows the derivatization mixture to be injected onto the HPLC column without need for excess reagent removal. Since the initial publication, a number of papers have described various applications and extensions of the original method *(7–11)*.

Fluorescence detection permits highly sensitive analyses with detection limits ranging from 50–300 fmol for the normal hydrolyzate amino acids. As little as 50 ng of protein hydrolyzate is sufficient for compositional analysis. Analysis of samples such as proteins recovered from 2-D gel electrophoresis, with typical spot amounts ranging from 100–100 ng, can be readily analyzed *(12)*.

2. Materials

2.1. Hydrolysis

1. 6×50 mm test tubes. Use borosilicate only.
2. Pyrrolysis oven (500°C capability) for glass cleaning (*see* **Note 1**).
3. Hydrolysis system for vapor phase hydrolysis.
4. Constant boiling HCl.
5. Phenol (crystalline or liquefied).

2.2. Derivatization

1. AQC dry powder. Store in a dry place, up to 6 mo in a desiccator. Available from Waters Corporation.
2. Borate derivatization buffer: $0.2\ M$ sodium borate, pH 8.8, with 5 m*M* calcium disodium ethylenediaminetetracetic acid (EDTA). Weigh 1.24 g of boric acid in a clean beaker. Add 100 mL of water, stir to dissolve. Add 187 mg of calcium disodium EDTA. Titrate to pH 8.8 with sodium hydroxide (a solution made fresh from pellets). Use the best quality boric acid and EDTA available.
3. Acetonitrile for reagent dissolution. Only use the highest quality with low water and alcohol content. AQC dry powder, borate derivatization buffer, and acetonitrile are available as a kit from Waters Corporation.
4. Reagent Solution: Dissolve the AQC powder in pure, dry acetonitrile to provide a solution that is approx 3.0 mg per mL. Heat at 50°C for up to 10 min to complete solubilization. Store in a desiccator for up to 1 wk (*see* **Notes 2** and **3**)
5. Amino acid calibration stock solution (2.5 m*M* in $0.1\ M$ HCl). Available from Pierce Chemical Co. or Sigma Chemical Co.
6. Working calibration mixture: Mix 40 µL of amino acid stock solution (2.5 m*M*) with 960 µL of water to give a solution with 100 pmol/µL of each AA
7. 20 m*M* HCl: Mix 10 µL of constant boiling HCl with 3 mL of high-performance liquid chromatography (HPLC)-grade water.

2.3. Chromatographic Analysis

1. Sodium acetate trihydrate, HPLC grade.
2. Triethylamine, Pierce Sequenal grade or equivalent.
3. Calcium disodium EDTA.
4. Acetonitrile (HPLC grade).
5. Phosphoric acid, 85%, reagent grade.
6. Working phosphoric acid. 10% concentrated phosphoric acid in water.
7. Working EDTA solution. Weigh 100 mg calcium disodium EDTA and dissolve in 100 mL of water. Store at 4°C for up to 3 mo.
8. Eluent A: 140 mM sodium acetate, 17 mM triethylamine (TEA), pH 5.05, containing 1 mM calcium disodium EDTA. Weigh 19.04 g sodium acetate. Add 1 L water, stir. Bring pH to approx 5.5 with working phosphoric acid (this reduces the odor from the TEA and improves its solubility in the aqueous buffer). Add 1 mL working EDTA. Add 2.37 mL of TEA (1.72 g) (*see* **Note 4**). Titrate to pH 5.05 with working phosphoric acid. Filter the buffer through 0.22- or 0.45-μm filters (*see* **Notes 5** and **6**). Use the HPLC-grade reagents and water. Store excess TEA under a blanket of inert gas.
9. Eluent B for ternary system, HPLC-grade acetonitrile.
10. Eluent C for ternary system, HPLC-grade water.
11. Alternate Eluent B for high-pressure mixing system, 60% acetonitrile. Mix 600 mL of acetonitrile with 400 mL of HPLC-grade water (Note: measure these volumes separately).
12. AccQ-Tag reversed phase column, 3.9 × 150 mm (*see* **Note 7**).
13. Guard column (3.9 × 20 mm) packed with reversed phase packing, compatible with analytical column.
14. Ternary gradient HPLC system equipped with an in-line degasser or sparging system, including column heater, autosampler, and dual monochromator fluorescence detector (*see* **Notes 8–11**).

3. Methods

3.1. Hydrolysis

1. Use a pipet or syringe to place a protein or peptide sample in a 6 × 50-mm glass hydrolysis tube.
2. Use a vacuum system, a stream of warm nitrogen, or a freeze dryer to remove the solvent from the sample.
3. Place the tubes in a vacuum sealable vial (e.g., a Pico-Tag® vacuum vial, Waters Corporation). Add 200 μL of constant boiling HCl into the bottom of the vacuum vial.
4. Add a crystal of phenol or a drop of liquefied phenol to the bottom of the vial, and seal the vial after three alternate steps each of vacuum and nitrogen purging, finishing with a vacuum step.
5. Hydrolyze the sample for 22–24 h at 110–112°C or for 1 h at 150°C.
6. Cool the samples. Open the vial, and remove the tubes with a forceps. Wipe the outside of the tubes with a Kimwipe. Remove excess HCl (present as a film on the

Table 1
Gradient for Ternary Eluent System

Time	Flow rate	%A	%B	%C	Curve
Initial	1.0	100	0	0	*
0.50	1.0	99	1	0	6
18.0	1.0	95	5	0	6
19.0	1.0	91	9	0	6
29.5	1.0	83	17	0	6
33.0	1.0	0	60	40	11
36.0	1.0	100	0	0	11

inside of the tube) under vacuum (5 min is more than sufficient) or a stream of warm nitrogen (*see* **Note 12**). **Caution:** Use nontalc gloves when handling strong acid!

3.2. Derivatization

3.2.1. Calibration Standard

1. Place 10 µL of dilute standard in a 6 × 50-mm tube using a syringe or microdisposable pipettor.
2. Add 70 µL of borate buffer; mix.
3. Add 20 µL of AQC, vortex immediately after addition (*see* **Notes 13** and **14**).
4. Transfer the contents of the tube to a conical-shaped HPLC sample vial and cap tightly with a silicone-lined septum.
5. Heat the vial in a reaction block or oven for 10 min at 50°C (*see* **Note 15**).
6. Inject 5 µL = 50 pmol.

3.2.2. Hydrolyzed Samples

1. Add 20 µL of 20 mM HCl to the sample and vortex (*see* **Note 16**).
2. Add 60 µL of borate buffer and vortex.
3. Add 20 µL of AQC reagent solution; vortex again.
4. Transfer the contents of the tube to a conical-shaped HPLC sample vial and cap tightly with a silicone-lined septum.
5. Heat the vial in a reaction block or oven for 10 min at 50°C.
6. Inject up to 20 µL of sample (*see* **Note 17**).

3.3. Chromatographic Analysis

3.3.1. Instrument Settings

1. Set the column heater to 38°C.
2. Set the fluorescence-detector excitation wavelength at 248 nm and the emission wavelength at 395 nm.

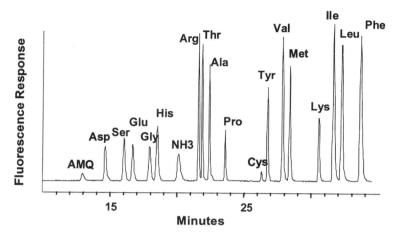

Fig. 1. Typical separation for a hydrolysate standard. The injected volume was 5 μL with a total of 50 pmol of each amino acid (25 pmol cystine) injected. Detection was accomplished by fluorescence with excitation at 248 nm and emission at 395 nm. The chromatographic conditions were those described in the methods section.

3.3.2. Chromatographic Operation

1. Equilibrate the system with 100% of eluent A for 10 min before beginning sample analysis.
2. Analyze samples using the following gradient shown in **Table 1**. The entire separation is completed in approx 35 min (*see* **Fig. 1**). Including column reequilibration, the total run-time for a typical HPLC system is 45 min (*see* **Notes 18–20**).

4. Notes

1. For high sensitivity analysis, it is necessary to have very clean tubes (e.g., fired at 500°C). Acid cleaning may be insufficient.
2. The AQC reagent can be slow to dissolve and may require brief sonication to solubilize. Make sure the vial is well sealed with a cap and further sealed with Parafilm, and limit the sonication to 60 s at a time. Be certain that no water gets in the reagent vial!
3. If the reconstituted reagent is kept dry, it should last for a minimum of several weeks. Repeated use from the same vial will expose the reagent to atmospheric water and slowly degrade the reagent. Minor hydrolysis will form AMQ plus CO_2 and NHS, normal byproducts of the derivatization procedure, and consequently will not contribute any significant interference. However, the AMQ produced will react slowly with another molecule of AQC, forming the symmetric *bis*-aminoquinoline urea. The urea has poor solubility and may form a precipitate

in the vial. Any urea in solution will elute after any derivatized amino acid and will also not interfere. However, continued breakdown will deplete the reagent, and unidentified fluorescent components that do interfere with the analysis do accumulate with continued use. For best results, store unused reagent in the following manner.

    a. Seal the vial as rapidly as possible following use.

    b. Wrap Parafilm® tightly around the top of the vial.

    c. Place the vial in a well-sealed desiccator.

    d. Although it has not been proven necessary, it may be useful to protect the vial from light with aluminum foil.

    e. If the vial is refrigerated or frozen, this may extend the life of the reagent. However it is imperative that the vial and the desiccator be warmed to room temperature before opening to prevent condensation of moisture. In addition, storage at subambient temperature often results in significant precipitation that may require sonication to resolubilize.

4. We have found it to be much more reproducible to weigh the TEA. Tare a clean vial with a cap. Pipet the approximate volume of TEA into the vial in a hood. Weigh the vial and add or remove the proper amount of TEA to correct the weight (density = 0.7255). Final weight should be within 0.05 g the desired amount. **Note:** To increase eluent A shelf-life, add 0.01% sodium azide. There will be no impact for fluorescence detection. UV-based systems will exhibit a gradually decreasing baseline. This can be offset by adding 100 μL of acetone to Eluent B *(7)*.

5. Poor quality eluents are a common source of background gradient contamination. This is especially problematic for operation with UV detection.

6. This eluent may be made as an x11 concentrate. Weigh 1904 g of sodium acetate. Add 1 L of water, stir thoroughly, and adjust the pH to 5.5 with phosphoric acid. Add 17.2 g of TEA and stir thoroughly. Add 10 mL of working EDTA solution. Titrate to pH 5.05 with phosphoric acid. To make working eluent, mix 100 mL of concentrate with 1000 mL of HPLC-grade water.

7. Other reversed phase columns can be substituted, but optimal separation may require substantial changes to the Eluent A composition, as well as the gradient profile.

8. It is advisable to use a small (e.g., 5 μL) flow cell in the detector. This will improve resolution compared to larger flow cells. Larger flow cells normally give a greater response.

9. Set the fluorescence emission bandwidth to the narrowest available setting. This will reduce the response from AMQ.

10. A UV detector set at 248 nm may be substituted for the fluorescence detector. Detection limits will be much higher, approx 1 pmol injected. There is no selectivity for detection. Thus, the AMQ peak is very large and can compromise the analysis of Asp. It may be necessary to adjust the gradient conditions and/or Eluent A pH to provide sufficient resolution *(7)*. Lowering Eluent A pH to 4.95 will improve the AMQ/Asp resolution. Decreasing the gradient slope between 0.5 and 18 min will also improve the resolution. There may be some deterioration in the resolution of Gly and His, or between Arg and Thr.

Table 2
Gradient for Binary Eluent System

Time	Flow rate	%A	%B	Curve
Initial	1.0	100	0	*
0.50	1.0	98	2	6
18.0	1.0	93	5	6
19.0	1.0	90	9	6
29.5	1.0	67	17	6
33.0	1.0	67	33	6
34.0	1.0	0	100	6
37.0	1.0	0	100	6
38	1.0	100	0	6

11. Other HPLC systems can be used. To use a high-pressure binary gradient, use the alternate Eluent B. A typical gradient is shown in **Table 2**. Use a total run-time of 50 min to equilibrate the column

12. Excess residual acid may lower the derivatization pH below the optimal range (8.2–10.0). One potential source is nonvolatile acid formation during the hydrolysis. Phosphate salt or sodium dodecyl sulfate will be converted to their respective acids, and will not be removed by vacuum drying. If excess acid is present, it may be neutralized with a TEA/ethanol/water solution (2:2:1, v/v). The volume should be kept to a minimum (10–20 µL, if possible) to avoid contamination problems. Vortex the dried sample with the TEA solution and vacuum dry the sample again. Alternatively, a higher concentration borate derivatization buffer (up to 1 M) may be used instead of the 0.2 M. The background from the derivatization blank may increase.

13. The amount of AQC added is sufficient to derivatize 5 µg of protein or peptide. Higher amounts may suffer lower yields for slow reacting amino acids, namely Asp, Glu, and Lys.

14. Less than quantitative derivatization of Lys results in production of either or both monoderivatized species. Poor derivatization, often caused by an imbalance of sample and reagent or poor pH control of the derivatization mixture, can be monitored by observing the appearance of these peaks small "satellite" peaks on the front shoulder of Gly and the trailing shoulder of His. Poor mixing of the reagent with the buffered sample may also result in less than quantitative recovery of fully derivatized Lys.

15. The purpose of the 10-min heating step is to convert the phenolic side chain of Tyr to free phenol and yield a product labeled only at the on the amino group. During the derivatization at room temperature, approx 30% of the phenol is labeled with AQC to form an unstable adduct. Heating the sample has no other significant effect.

16. The addition of HCl at this step helps improve the recovery of the basic amino acids Lys, His, and Arg.

17. Peak splitting for Asp and Ser and other early eluting components can be caused by excessive injection volume. To avoid this phenomenon, limit the injection volume to ≤ 20 µL.
18. Poor resolution between AMQ and Asp can be caused by too steep of an initial gradient slope or Eluent A pH being too high. Note that gradient delivery may vary with different HPLC systems.
19. Poor resolution of Asp and Ser or Glu and Gly can be caused by Eluent A pH being too low.
20. Poor resolution of Cys-Cys and Tyr can be caused by (1) column temperature too high; (2) gradient slope from 19–28 min too shallow; or (3) Eluent A pH too high.

References

1. Hill, D. W., Walters, F. H., Wilson, T. D., and Stuart, J. D. (1979) High performance liquid chromatographic determination of amino acids in the picomole range. *Anal. Chem.* **51,** 1338–1341.
2. Koop, D. R., Morgan, E. T., Tarr, G. E., and Coon, M. J. (1982) Purification and characterization of a unique isozyme of cytochrome P-450 from liver microsomes of ethanol-treated rabbits. *J. Biol. Chem.* **257,** 8472–8480.
3. Bidlingmeyer, B. A., Cohen, S. A., and Tarvin, T. L. (1984) Rapid analysis of amino acids using pre-column derivatization. *J. Chromatog.* **336,** 93–104.
4. Heinrikson, R. L. and Meredith, S. C. (1984) Amino acid analysis by reversed-phase high-performance liquid chromatography: pre-column derivatization with phenylisothiocyanate. *Anal. Biochem.* **136,** 65–74.
5. Cohen, S. A. and Strydom, D. J. (1988) Amino acid analysis utilizing phenylisothiocyanate derivatives. *Anal. Biochem.* **174,** 1–16.
6. Cohen, S. A. and Michaud, D. P. (1993) Synthesis of a fluorescent derivatizing reagent, 6-aminoquinolyl-*N*-hydroxysuccinimidyl carbamate, and its application for the analysis of hydrolysate amino acids via high performance liquid chromatography. *Anal. Biochem.* **211,** 279–287.
7. Liu, H. J. (1994) Determination of amino acids by pre-column derivatization with 6-aminoquinolyl-*N*-hydroxysuccinimidyl carbamate and high performance liquid chromatography with ultraviolet detection. *J. Chromatog.* **670,** 59–66.
8. Strydom, D. J. and Cohen, S. A. (1994) Comparison of amino acid analyses by phenylisothiocyanate and 6-aminoquinolyl-*N*-hydroxysuccinimidyl carbamate pre-column derivatization. *Anal. Biochem.* **222,** 19–28.
9. Strydom, D. J. and Cohen, S. A. (1993) Sensitive analysis of cystine/cysteine using 6-aminoquinolyl-*N*-hydroxysuccinimidyl carbamate derivatives, in *Techniques in Protein Chemistry IV* (Angeletti, R. H. ed.), Academic, San Diego, CA, pp. 299–306.
10. Cohen, S. A. and De Antonis, K. M. (1994) Applications of amino acid analysis derivatization with 6-aminoquinolyl-*N*-hydroxysuccinimidyl carbamate: Analysis of feed grains, intravenous solutions and glycoproteins. *J. Chromatog.* **661,** 25–34.
11. Van Wandelen, C. and Cohen, S. A. (1997) Using quaternary high-performance liquid chromatography eluent systems for separating 6-aminoquinolyl-*N*-

hydroxysuccinimidyl carbamate-derivatized amino acid mixtures. *J. Chromatog.* **763,** 11–22.

12. Grant, R. A. and Fieno, A. M. (1998) Use of the AQC Pre-column method for determination of protein composition and identification from PVDF blots. Presented at ABRF'98: From Genomes to Function — Technical Challenges of the Post-Genome Era, http://www. abrf. org/ABRF/ABRFMeetings/abrf98/rgrant. pdf.

5

Amino Acid Analysis by High-Performance Liquid Chromatography after Derivatization with 1-Fluoro-2,4-dinitrophenyl-5-L-alanine Amide (Marfey's Reagent)

Sunil Kochhar, Barbara Mouratou, and Philipp Christen

1. Introduction

Precolumn derivatization of amino acids followed by their resolution by reverse-phase high-performance liquid chromatography (RP-HPLC) is now the preferred method for quantitative amino acid analysis. The derivatization step introduces covalently bound chromophores necessary not only for interactions with the apolar stationary phase for high resolution, but also for photometric or fluorometric detection.

Marfey's reagent, 1-fluoro-2,4-dinitrophenyl-5-L-alanine amide or (S)-2-[(5-fluoro-2,4-dinitrophenyl)-amino]propanamide, can be used to separate and to determine enantiomeric amino acids (1). The reagent reacts stoichiometrically with the amino group of enantiomeric amino acids to produce stable diastereomeric derivatives that can readily be separated by RP-HPLC (**Fig. 1**). The dinitrophenyl alanine amide moiety strongly absorbs at 340 nm ($\varepsilon = 30,000$ M^{-1} cm^{-1}), allowing detection in the subnmol range.

Precolumn derivatization with the reagent is also used to quantify the 19 commonly analyzed L-amino acids (2). Major advantages of Marfey's reagent over other precolumn derivatizations are: (1) possibility to carry out chromatography on any multipurpose HPLC instrument without column heating; (2) detection at 340 nm is insensitive to most solvent impurities; (3) simultaneous detection of proline in a single chromatographic run; and (4) stable amino acid derivatives. For the occasional user, the simple methodology provides an attractive and inexpensive alternative to the dedicated amino acid analyzer. The

From: *Methods in Molecular Biology*, vol. 159: *Amino Acid Analysis Protocols*
Edited by: C. Cooper, N. Packer, and K. Williams © Humana Press Inc., Totowa, NJ

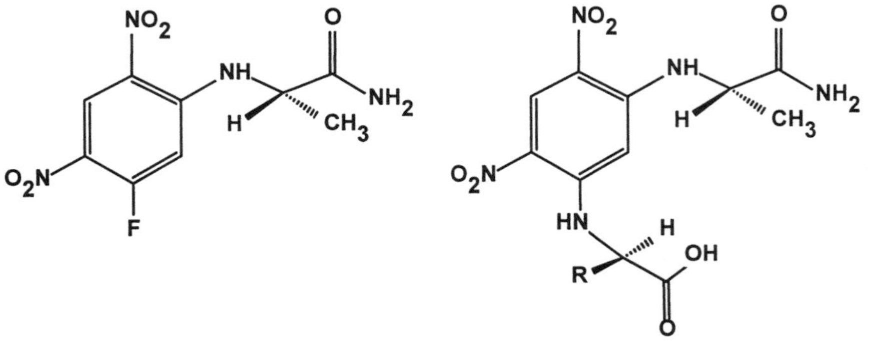

Fig. 1. Marfey's reagent (I) and L,L-diastereomer derivative from L-amino acid and the reagent (II).

precolumn derivatization with Marfey's reagent has found applications in many diverse areas of biochemical research (3–14) including determination of substrates and products in enzymic reactions of amino acids (see **Note 1**).

2. Materials
2.1. Vapor-Phase Protein Hydrolysis

1. Pyrex glass vials (25–50 × 5 mm) from Corning.
2. Screw-cap glass vials.
3. 6 N HCl from Pierce.

2.2. Derivatization Reaction

1. Amino acid standard solution H from Pierce.
2. L-Amino acid standard kit LAA21 from Sigma.
3. Marfey's reagent from Pierce. **Caution:** Marfey's reagent is a derivative of 1-fluoro-2,4-dinitrobenzene, a suspected carcinogen. Recommended precautions should be followed in its handling (15).
4. HPLC/spectroscopic grade triethylamine, methanol, acetone, and dimethyl sulfoxide (DMSO) from Fluka.

2.3. Chromatographic Analysis

1. Solvent delivery system: Any typical HPLC system available from a number of manufacturers can be used for resolution of derivatized amino acids. We have recently used HPLC systems from Bio-Rad (HRLC 800 equipped with an autoinjector AST 100), Hewlett Packard 1050 system and Waters system.
2. Column: A silica-based C_8 reverse-phase column, e.g., LiChrospher 100 RP 8 (250 × 4.6 mm; 5 µm from Macherey-Nagel), Aquapore RP 300 (220 × 4.6 mm;

7 μm from Perkin-Elmer), Nucleosil 100-C_8 (250 × 4.6 mm; 5 μm from Macherey-Nagel), or Vydac C_8 (250 × 4.6 mm; 5 μm from The Sep/a/ra/tions group).
3. Detector: A UV/vis HPLC detector equipped with a flow cell of lightpath 0.5–1 cm and a total volume of 3–10 μL (e.g., Bio-Rad 1790 UV/VIS monitor, Hewlett Packard Photodiode Array 1050, Waters Photodiode Array 996). The derivatized amino acids are detected at 340 nm.
4. Peak Integration: Standard PC-based HPLC software with data analysis program can be employed to integrate and quantify the amino acid peaks (e.g., ValueChrom from Bio-Rad, Chemstation from Hewlett Packard or Millennium from Waters).
5. Solvents: The solvents should be prepared with HPLC-grade water and degassed.
   a. Solvent A: 13 m*M* trifluoroacetic acid (TFA) plus 4% (v/v) tetrahydrofuran in water.
   b. Solvent B: Acetonitrile (50% v/v) in solvent A.

3. Methods
3.1. Vapor-Phase Protein Hydrolysis

1. Transfer 50–100 pmol of protein sample or 20 μL of amino acid standard solution H containing 2.5 μmol/mL each of 17 amino acids into glass vials, and dry under reduced pressure in a SpeedVac concentrator.
2. Place the sample vials in a screw-cap glass vial containing 200 μL of 6 *N* HCl.
3. Flush the vial with argon for 5–15 min, and cap it air tight.
4. Incubate the glass vial at 110°C for 24 h or 150°C for 2 h in a dry-block heater.
5. After hydrolysis, remove the glass vials from the heater, cool to room temperature, and open it slowly.
6. Remove the insert vials, wipe their outside clean with a soft tissue paper, and dry under reduced pressure.

3.2. Derivatization Reaction

1. Add 50 μL of triethylamine/methanol/water (1:1:2) to the dried sample vials, mix vigorously by vortexing, and dry them under reduced pressure (*see* **Note 2**).
2. Prepare derivatization reagent solution (18.4 m*M*) by dissolving 5 mg of Marfey's reagent in 1 mL of acetone.
3. Dissolve the dried amino acid mixture or the hydrolyzed protein sample in 100 μL of 25% (v/v) triethylamine, add 100 μL of the Marfey's reagent solution, and mix by vortexing (*see* **Note 3**).
4. Incubate the reaction vial at 40°C for 60 min with gentle shaking protected from light (*see* **Note 4**).
5. After incubation, stop the reaction by adding 20 μL of 2 *N* HCl (*see* **Note 5**). Dry the reaction mixture under reduced pressure.
6. Store the dried samples at –20°C in dark until used (*see* **Note 6**).
7. Dissolve the dry sample in 1 mL of 50% (v/v) DMSO.

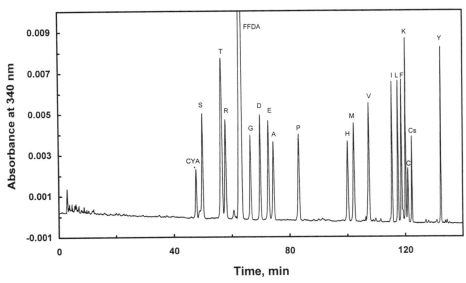

Fig. 2. Separation of 2,4-dinitrophenyl-5-L-alanine amide derivatives of L-amino acids by HPLC. A 20-µL aliquot from the amino acid standard mixture (standard H from Pierce Chemical Company) was derivatized with Marfey's reagent. Chromatographic conditions: Column, LiChrospher 100 RP 8 (250 × 4.6 mm; 5 µm from Macherey-Nagel); sample, 100 pmol diastereomeric derivatives of 18 L-amino acids; solvent A, 13 m*M* trifluoroacetic acid plus 4% (v/v) tetrahydrofuran in water; solvent B, 50% (v/v) acetonitrile in solvent A; flow rate, 1 mL/min; detection at 340 nm; elution with a linear gradient of solvent B in solvent A (**Table 1**). The amino acid derivatives are denoted by single-letter code, cysteic acid as CYA, and cystine as Cs. FFDA indicates the reagent peak; the peaks without denotations are reagent-related, as shown in an independent chromatographic run of the reagent alone.

3.3. Chromatographic Analysis

1. Prime the HPLC system according to the manufacturer's instructions with solvent A and solvent B.
2. Equilibrate the column and detector with 90% solvent A and 10% solvent B.
3. Bring samples for analysis to room temperature; dilute, if necessary, with solvent A, and inject 20 µL (50–1000 pmol) onto the column (*see* **Note 7**).
4. Elute with gradient as described in **Table 1** (*see* **Note 8**). Resolution of 2,4-dinitrophenyl-5-L-alanine amide derivatives of 18 commonly occurring L-amino acids and of cysteic acid is achieved within 120 min (**Fig. 2**) (*see* **Note 9**).
5. Determine the response factor for each amino acid from the average peak area of standard amino acid chromatograms at 100, 250, 500, and 1000 pmol amounts.
6. When HPLC is completed, wash the column and fill the pumps with 20% (v/v) degassed methanol for storage of the system. Before reuse, purge with 100% methanol.

Table 1
HPLC Program for Resolution of Amino Acids Derivatized
with Marfey's Reagent

Time (min)	% Solvent A	% Solvent B	Input
0	90	10	
15	90	10	Detector auto zero
15.1	90	10	Inject
115	50	50	
165	0	100	
170	0	100	

4. Notes

1. For analysis of amino acids as substrates or products in an enzymic reaction, deproteinization is required. Add 4 M perchloric acid to a final concentration of 1 M and incubate on ice for at least 15 min. Excess perchloric acid is precipitated as $KClO_4$ by adding an equal volume of ice-cold 2 M KOH. After centrifugation for 15 min at 4°C, the supernatant is collected, dried and used for derivatization.
2. Complete removal of HCl is absolutely essential for quantitative reaction between amino acids and Marfey's reagent.
3. The molar ratio of the Marfey's reagent to the total amino acids should not be more than 3:1.
4. During derivatization, the color of the reaction mixture turns from yellow to orange–red.
5. On acidification, the color of the reaction mixture turns to yellow.
6. The dried amino acids are stable for over one month when stored at –20°C in the dark. In 50% (v/v) DMSO, derivatives are stable for 72 h at 4°C and for over 6 wk at –20°C.
7. To obtain good reproducibility, an HPLC autoinjector is highly recommended.
8. Silica-based C_8 columns from different commercial sources produce base-line resolution of the 19 amino acid derivatives; nevertheless, the gradient slope of solvent B should be optimized for each column. *S*-Carboxymethyl-L-cysteine and tryptophan, if included, are also separated in a single chromatographic run (*see* **ref. 2**).
9. The identity of each peak is established by adding a threefold molar excess of the amino acid in question. A reagent blank should be run with each batch of the Marfey's reagent to identify reagent-related peaks. Excess reagent interferes with base-line separation of the arginine and glycine peaks. Lysine and tyrosine are separated as disubstituted derivatives.

References

1. Marfey, P. (1984) Determination of D-amino acids. II. Use of a bifunctional reagent, 1, 5-difluoro-2, 4-dinitrobenzene. *Carlsberg Res. Commun.* **49,** 591–596.

2. Kochhar, S. and Christen, P. (1989) Amino acid analysis by high-performance liquid chromatography after derivatization with 1-fluoro-2, 4-dinitrophenyl-5-L-alanine amide. *Anal. Biochem.* **178,** 17–21.

3. Kochhar, S. and Christen, P. (1988) The enantiomeric error frequency of aspartate aminotransferase. *Eur. J. Biochem.* **175,** 433–438.

4. Szókán, G., Mezö, G., and Hudecz, F. (1988) Application of Marfey's reagent in racemization studies of amino acids and peptides. *J. Chromatog.* **444,** 115–122.

5. Martínez del Pozo, A., Merola, M., Ueno, H., Manning, J. M., Tanizawa, K., Nishimura, K., et al. (1989) Stereospecificity of reactions catalyzed by bacterial D-amino acid transaminase. *J. Biol. Chem.* **264,** 17784–17789.

6. Szókán, G., Mezö, G., Hudecz, F., Majer, Z., Schön, I., Nyéki, O., et al. (1989) Racemization analyses of peptides and amino acid derivatives by chromatography with pre-column derivatization. *J. Liq. Chromatogr.* **12,** 2855–2875.

7. Adamson, J. G., Hoang, T., Crivici, A., and Lajoie, G. A. (1992) Use of Marfey's reagent to quantitate racemization upon anchoring of amino acids to solid supports for peptide synthesis. *Anal. Biochem.* **202,** 210–214.

8. Goodlett, D. R., Abuaf, P. A., Savage, P. A., Kowalski, K. A., Mukherjee, T. K., Tolan, J. W., et al. (1995) Peptide chiral purity determination: hydrolysis in deuterated acid, derivatization with Marfey's reagent and analysis using high-performance liquid chromatography-electrospray ionization-mass spectrometry. *J. Chromatog.* A **707,** 233–244.

9. Graber, R., Kasper, P., Malashkevich, V. N., Sandmeier, E., Berger, P., Gehring, H., et al. (1995) Changing the reaction specificity of a pyridoxal-5'-phosphate-dependent enzyme. *Eur. J. Biochem.* **232,** 686–690.

10. Vacca, R. A., Giannattasio, S., Graber, R., Sandmeier, E., Marra, E., and Christen, P. (1997) Active-site Arg→Lys substitutions alter reaction and substrate specificity of aspartate aminotransferase. *J. Biol. Chem.* **272,** 21932–21937.

11. Wu, G. and Furlanut, M. (1997) Separation of DL-dopa by means of micellar electrokinetic capillary chromatography after derivatization with Marfey's reagent. *Pharmacol. Res.* **35,** 553–556.

12. Goodnough, D. B., Lutz, M. P., and Wood, P. L. (1995) Separation and quantification of D- and L-phosphoserine in rat brain using *N*-alpha-(2, 4-dinitro-5-fluorophenyl)-L-alaninamide (Marfey's reagent) by high-performance liquid chromatography with ultraviolet detection. *J. Chromatog. (B. Biomed. Appl.)* **672,** 290–294.

13. Yoshino, K., Takao, T., Suhara, M., Kitai, T., Hori, H., Nomura, K., et al. (1991) Identification of a novel amino acid, o-bromo-L-phenylalanine, in egg-associated peptides that activate spermatozoa. *Biochemistry* **30,** 6203–6209.

14. Tran, A. D., Blanc, T., and Leopold, E. J. (1990) Free solution capillary electrophoresis and micellar electrokinetic resolution of amino acid enantiomers and peptide isomers with L- and D-Marfey's reagents. *J. Chromatog.* **516,** 241–249.

15. Thompson, J. S. and Edmonds, O. P. (1980) Safety aspects of handling the potent allergen FDNB. *Ann. Occup. Hyg.* **23,** 27–33.

6

The Analysis of Amino Acids Using Precolumn Derivatization, HPLC, and Electrochemical Detection

C. David Forster and Charles A. Marsden

1. Introduction

The amino acids we have the most experience with are GABA, glutamate and arginine, though other amino acids can be measured. The method used is based on a derivatization reported by Jacobs (1) using o-phthalaldehyde in the presence of sulphite. The method was used by Pearson et al. (2), Smith and Sharp (3), and was further developed by Rowley et al. (4).

The amino acid content of brain microdialysates can be analyzed using this method as described (4). Other matrices can also be analyzed, but it is necessary to incorporate a sample purification step. It should be noted that although the derivatized amino acids can be analyzed by electrochemical detection, the derivatives are also fluorophors and can therefore be analyzed using fluorescence detection. However, electrochemical detection offers an extra degree of specificity because of the need for molecules to be electroactive.

Electrochemical detection (ECD) when linked with high-performance liquid chromatography (HPLC) is a very selective and sensitive analytical tool. It is dependent on the analyte of interest being electroactive, i.e., the molecule will either be oxidized or reduced when an electric potential is applied to it. Most ECD detectors are made up of a working electrode (to which a set potential is applied), a reference electrode, and an auxiliary electrode (which is used to keep the potential within the cell constant). The technique was first described by Shoup and Kissinger (5). Since then, it has been developed and used to analyze many different molecules. There are two main configurations of electrode; the cross flow where the column eluent passes over the working elec-

From: *Methods in Molecular Biology*, vol. 159: *Amino Acid Analysis Protocols*
Edited by: C. Cooper, N. Packer, and K. Williams © Humana Press Inc., Totowa, NJ

trode and electrolysis takes place at the solid liquid interface or the wall jet in which the eluent is directed at the electrode surface. Working electrode surfaces can be carbon paste, gold, carbon/gold, nickel, silver, platinum, or glassy carbon. The working electrode used for amino acids is the glassy carbon type, in both the cross flow or the wall jet cells. The reference electrode is Ag/AgCl.

The two makes of electrode we use are the BAS (Bioanalytical Supplies) cross flow or the Antec Leyden (Leyden) wall jet. Both operate on the same basic electrochemical principles, but the day-to-day operation is subtly different, and we recommend study of the relevant instruction book (*see* **Note 1**).

2. Materials
2.1. Equipment

1. The HPLC system is of utmost importance. The pump must provide a pulse-free flow and it is essential to incorporate a pulse-damping device. The pulse damper we use, and have used successfully, on different makes of pump is supplied by Presearch (Hitchin). Our system uses a Gynkotek Model 300 pump (the flow rate used was 0.3 mL min^{-1}–1 mL min^{-1} dependent on peak of interest); samples were injected using a Rheodyne 9125 valve with a 20-μL loop. The electrochemical detector used was supplied by Antec Leyden (the detector potential used was 0.7–0.85 V dependent on amino acid of interest), and it is linked to a Spectra Physics SP 4290 data collection system.
2. Column: Rainin Dynamax C18 5 μm, 25 cm × 4.6 mm.
            Spherisorb ODS 5 μm, 25 cm × 4.6 mm.

2.2. Reagents

All chemicals are HPLC grade or better. We used OPA (o-phthalaldehyde) from Sigma (Poole), this provided the cleanest reagent blank (*see* **Note 2**). The water used is double-distilled deionized quality, which is then again purified using an ELGA UHQ water purification system prior to use.

2.3. Solutions

1. Solution 1: 1 M sodium sulphite in water (*see* **Note 3**).
2. Solution 2: 0.1 M sodium tetraborate in water. Adjusted to pH 10.4 with 5 M sodium hydroxide.
3. Solution 3: Mix 0.5 mL solution 1 and 0.5 mL absolute ethanol with 9 mL solution 2.
4. Derivatization solution: Weigh 22 mg OPA and add solution 3 (10 mL) (*see* **Notes 4** and **5**).
5. Mobile phase: 0.1 M monosodium phosphate, 0.5 mM ethylenediaminetetracetic acid (EDTA), Methanol, Water. Adjusted to pH 4.5 with 1 M phosphoric acid. The percentage methanol in the mobile phase should be adjusted to give the best separation for the peak or peaks of interest (*see* **Note 6**).

3. Methods

3.1. Amino Acid Content of Standard Solutions

Add 20 μL of derivatization solution to 1 mL of amino acid standard solution (*see* **Notes 7** and **8**) and allow to react for 10 min at room temperature before injection onto the HPLC column. **Figure 1** shows the chromatogram produced by 20 pmol derivatized amino acid mixed standard. The standards are prepared from a 0.01 *M* stock solution to give a range of concentrations equivalent to 0–12 pmol injected onto the column. **Figure 2** shows the calibration curve obtained. We have found 10 min to be the optimum reaction time (*see* **Note 9**) *(4)* (**Fig. 3**).

3.2. Amino Acid Content Of Dialysates

Collect brain microdialysis samples and rapidly freeze using liquid nitrogen. They are then stored at –80°C until analysis; this is usually within 1 mo of the sample being taken.

On the day of analysis the following procedure is followed.

1. Prepare derivatization solution as listed above. Store at room temperature in a closed vial in the dark (a drawer is suitable).
2. Prepare standard solutions from the stock, and store in the refrigerator until required.
3. Add 20 μL of OPA solution to an Eppendorf tube (or equivalent plastic tube). Add 1 mL of standard solution to the OPA, and allow to stand for 10 min at room temperature. Inject the resultant mixture onto the HPLC column. It may be necessary to microfilter before injection. Syringe filters are available from many suppliers, tests should be carried out to ensure that there is no retention of the analytes on the filter caused by interactions with the filter material. Retention time will be dependent on the amino acid of interest, please refer to the chromatogram as a reference to the retention times (**Fig. 3**).

   It is well known that there are some differences in column performance between manufacturers, so the retention times may alter using different commercial column packings, although it is C18.
4. When the standard chromatogram has finished, add 20 μL of dialysate solution to 0.4 μL of OPA mixture and allowed to react for 10 min at room temperature before injection onto the HPLC column. (**Fig. 3**) shows the chromatogram produced by 20 μL of dialysate.

3.3. Amino Acid Content of Human Seminal Plasma

1. Add 25 μL of absolute ethanol (*see* **Note 10**) to 50 μL of seminal fluid, and store the resultant mixture at –80°C until assayed.
2. Defrost the sample, add 25 μL of distilled water, and centrifuge the mixture for 10 min to separate the protein pellet.

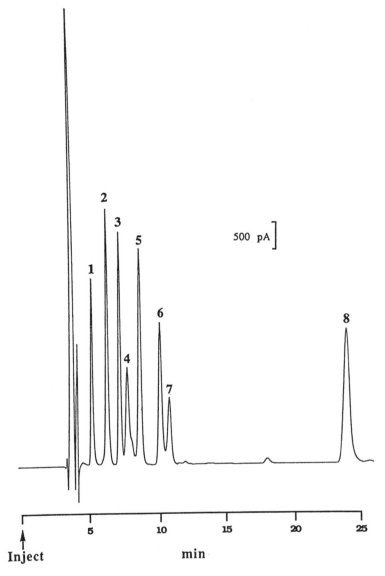

Fig. 1. The chromatogram of a mixed standard of derivatized amino acids equivalent to 20 pmol on column. Peak 1, serine; 2, glycine; 3, taurine; 4, glutamate; 5, arganine; 7, OPA; 8, GABA.

3. Remove 20 µL of supernatant and add to 0.4 µL of derivatizing agent and 180 µL of distilled water (*see* **Note 7**). Allow the resultant mixture to react for 10 min at room temperature before injecting 20 µL onto the HPLC column.

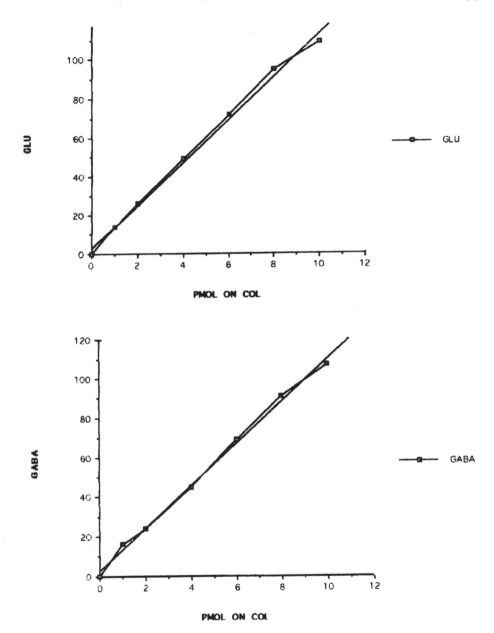

Fig. 2. The calibration curve of 0–12 pmol of derivatized GABA and glutamate.

4. Notes

1. Particular attention must be paid to the care of the electrode and its surface, which is essential for successful operation.

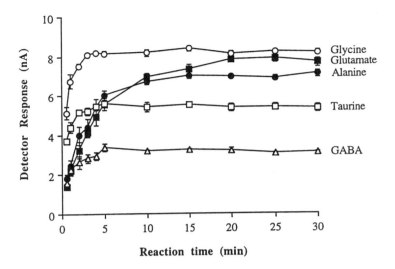

Fig. 3. The detector response of amino acid standards as a function of derivatizing time.

2. If OPA is obtained from other suppliers, then a reagent blank experiment must be carried out to determine the purity of the OPA as this is essential for high-quality baselines, we have experienced variability in quality between supplies. It may be possible to use other C18 columns for this analysis. We have not evaluated any other columns for this separation.

3. The use of sulphite as the intermediary in the derivatization avoids the stench associated with thiols.

4. We have found it easier to add solution 3 to the OPA, rather than adding each component of the mixture separately.

5. The OPA should be kept at room temperature in the dark. Fresh OPA solution should be prepared daily. There is some precipitation if the mixture is left in a refrigerator.

6. **ECD is a destructive analytical method.** Therefore, the mobile phase can be recirculated, but if you want to use another detector in series, ensure the nonde-structive detector is first in the mobile phase stream. Likewise if collection for further analysis is required, it is best to collect without the detector in line. To do this, a chromatogram of standards must be obtained to localize the peaks, collection is then done under the same chromatographic conditions at the retention times estimated.

7. To economize on disposable pipet tips, we add the derivatizing solution to the empty tube before the sample so that contamination of the tip does not occur and one tip can be reused.

8. Use plastic tubes for the reactions to avoid the amino acid sticking to the vessel walls.

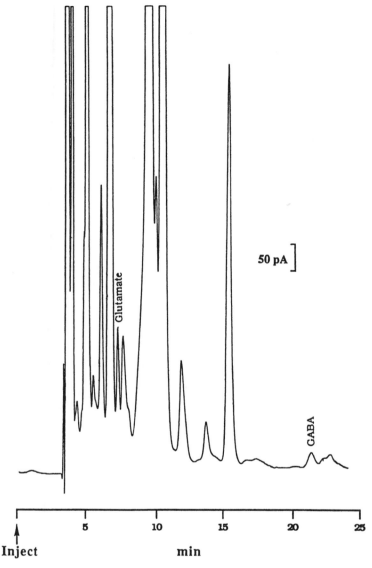

Fig. 4. The chromatogram produced by 20 μL derivatized dialysate. Glutamate and GABA are indicated.

9. This timing has been found to give maximal conversion for GABA, glutamate, and arginine (**Fig. 4**). It would be advisable to conduct a reaction-kinetic experiment to confirm the optimal conditions for the derivatization of the amino acids you wish to measure. It should be noted that this reaction timing was developed in England, and it could be that your lab is either warmer (for those of you lucky

enough to have a better climate than ours) or colder (for those of you lucky enough to have better air-conditioning than ours) and this may alter the reaction kinetics. We found the resultant derivatized amino acids to be stable at room temperature for up to 5 h, however, stability studies should be conducted to confirm the local environment does not have any effect.

10. Ethanol is used as a protein precipitator as the derivatization mixture contains ethanol.

11. As experience is gained, it is possible then to time the procedures so that a derivatization is running at the same time as a chromatogram. The procedure is labor intensive and the operator has much time when nothing else productive can be started without interruption. There are some autoinjectors on the market, which can be programmed to take some of the repetitiveness out of the assay. However, it should be noted that the OPA solution only has a short shelf life. If the autoinjector can carry out the derivatization procedure automatically, then this should not cause a problem. If derivatization is carried out manually before autoinjection, then the 5-h stability must be noted and the batch tailored to this timing. We use a Triathlon (Spark Holland) autoinjector with peltier cooling for another procedure but it can be adapted for derivatization.

12. All aspects of laboratory safety must be adhered to, i.e., laboratory coats, safety spectacles, gloves, and anything required in local regulations. It should be remembered that HPLC uses high-pressure liquids; if a joint fractures, it is possible for the high-pressure liquid to be pushed through the dermis.

13. The local rules within the School of Biomedical Sciences, Nottingham University, U. K., stress that all workers who have potential contact with human body fluids must have a resistance to hepatitis B. If the worker does not show resistance, then a course of hepatitis B immunization must be undertaken before any work is started.

References

1. Jacobs W. A. (1987) o-Phthaldehyde-sulfite derivatisation of primary amines for liquid chromatography-electrochemistry. *J. Chromatog.* **392,** 435–441.
2. Pearson S. J., Czudek C., Mercer K., and Reynolds G. P. Electrochemical detection of human brain transmitter amino acids by high perfomance liquid chromatography of stable o-phthalaldehyde-sulphite derivatives. *J. Neural Transm.* **86,** 151–157.
3. Smith, S. and Sharp T, in vivo measurement of extracellular GABA in rat nucleus accumbens: improved methodology. *Br. J. Pharmacol.* **107,** 210P
4. Rowley H. L., Martin K. F., and Marsden C. A. (1995) Determination of in vivo amino acid neurotransmitters by high-performance liquid chromatography with o-phthalaldehyde-sulphite derivatisation and electrochemical detection. *J. Neurosci. Methods* **57,** 93–99.
5. Shoup R. E. and Kissinger P. T. (1976) A versatile thin layer detector cell for high performance liquid chromatography. *Chem. Instrument.* **7(3),** 171–177.

7

Anion Exchange Chromatography and Integrated Amperometric Detection of Amino Acids

Petr Jandik, Christopher Pohl, Victor Barreto,
and Nebojsa Avdalovic

1. Introduction

During the 1980s and 1990s, electrochemical detection has gained acceptance as a method of choice for some important biomolecules. Initially, the technique was utilized with glassy carbon electrodes and in the mode of constant potential amperometry. The first widely used applications were for catecholamines and their derivatives *(1)*. Today, most laboratories around the world still rely on electrochemical detection when it comes to analyzing catecholamines in biological fluids. Although that approach has been highly successful, the next important advance came in the form of pulsed amperometric detection (PAD). The introduction of voltage pulses broadens the scope of electrochemical detection by regenerating the electrode in time for each new detection measurement. Gradual or sudden fouling of the electrode surface by the products of electrode reaction and the corresponding decreased detection response is eliminated with properly designed PAD techniques. The first important application of PAD was for carbohydrates following their separation by anion exchange *(2,3)*.

The analysis of amino acids based on anion exchange chromatography and integrated pulsed amperometric detection (IPAD) has evolved from Johnson and Polta's *(4)* original approach reported in 1983. The first high-performance liquid chromatography (HPLC) system based on that methodology was introduced in 1999 *(5)*.

1.1. Anion Exchange of Amino Acids

Alkaline pH is a requirement for the IPAD detection of amino acids on gold electrodes. Alkaline pH of anion exchange eluents thus makes IPAD possible.

From: *Methods in Molecular Biology*, vol. 159: *Amino Acid Analysis Protocols*
Edited by: C. Cooper, N. Packer, and K. Williams © Humana Press Inc., Totowa, NJ

Table 1
Amino Acids and Other Compounds Analyzed by AAA Direct

Compound	Type	Compound	Type
1 Arginine	Amino acid	29 Glycine	Amino acid
2 *Mannitol*	*Alditol*	30 *Ribose*	*Monosaccharide*
3 *Fucose*	*Monosaccharide*	31 Valine	Amino acid
4 Hydroxy-Lysine	Hydroxy-amino acid	32 Hydroxy-Proline	Hydroxy-amino acid
5 *Gly-Lys*	*Peptide*	33 *Gly-OHPro*	*Peptide*
6 Ornithine	Amino acid	34 Serine	Amino acid
7 *MetSulfoxide*	*Ox. Product of Met*	35 *Sucrose*	*Disaccharide*
8 Lysine	Amino acid	36 *Gly-Ser*	*Peptide*
9 *Gly-Pro*	*Peptide*	37 Proline	Amino acid
10 *Rhamnose*	*Monosaccharide*	38 *Isomaltose*	*Disaccharide*
11 Glucosamine	Amino-sugar	39 IsoLeucine	Amino acid
12 Galactosamine	Amino-sugar	40 Leucine	Amino acid
13 *Arabinose*	*Monosaccharide*	41 Methionine	Amino acid
14 Glutamine	Amino acid	42 NorLeucine	Amino acid
15 Citrulline	Amino acid	43 Taurine	Amino sulfonic acid
16 *Mannose*	*Monosaccharide*	44 Histidine	Amino acid
17 α Amino-butyric	Amino acid	45 *Phospho-Arginine*	*Phospho-amino acid*
18 Asparagine	Amino acid	46 Diaminopimelic	Amino acid
19 *Glucose*	*Monosaccharide*	47 Phenylalanine	Amino acid
20 *Galactose*	*Monosaccharide*	48 Glutamate	Amino acid
21 *Xylose*	*Monosaccharide*	49 Aspartate	Amino acid
22 *(Gly)3*	*Peptide*	50 Cystine	Amino acid
23 Alanine	Amino acid	51 *Phospho-Serine*	*Phospho-amino acid*
24 *Gly-Gly*	*Peptide*	52 *Phospho-Threonine*	*Phospho-amino acid*
25 Threonine	Amino acid	53 Tyrosine	Amino acid
26 *(Gly)4*	*Peptide*	54 *Phospho-Tyrosine*	*Phospho-amino acid*
27 γAmino-butyric	Amino acid	55 *Cysteic Acid*	*Ox. Prod. of Cys/(Cys)2*
28 *Methionine Sulfone*	*Ox. Product of Met*	56 Tryptophan	Amino acid

The italics denote compounds usually separated by modified gradients. The actual elution order for these compounds may vary depending on the gradient method selected. **Note:** Separation of methionine oxidation products requires a different column temperature.

Table 2
Gradient Conditions for Analyzing Amino Acids in Hydrolysates

Time (min)	%E1	%E2	%E3	Curve	Comments (Flow rate: 0.25 mL/min)
Init	76	24	0		Autosampler fills the sample loop
0.00	76	24	0		Valve from Load to Inject
2	76	24	0		Begin OH gradient, valve from Inject to Load
8	64	36	0	8	
11	64	36	0		Begin acetate gradient
18	40	20	40	8	
21	44	16	40	5	
23	14	16	70	8	
42	14	16	70		
42.1	20	80	0	5	Column wash with hydroxide
44.1	20	80	0		
44.2	76	24	0	5	Start of reequilibration to initial conditions
75	76	24	0		

As expected, anion exchange causes the amino acids to elute in a reverse order to that observed on the more commonly used cation exchangers *(6)*. **Table 1** lists a selection of amino acids and other compounds that can be separated by the anion exchange method. All the separations are achieved by ternary gradients employing three simple eluents (E1: water, E2: 0.25 M NaOH, and E3: 1.0 M NaAc). The three most frequently utilized gradient programs are listed in **Tables 2–4**. The corresponding chromatograms of standard mixtures are shown in **Figs. 1–3**.

1.2. Principle of Integrated Pulsed Amperometric Detection

With IPAD, one can achieve sensitivities comparable to or better than those obtained by detecting fluorescence of various amino acid derivatives. The most important advantage of IPAD is that it does not require any form of derivatization to make the analytes detectable. As illustrated in **Figs. 2** and **3**, IPAD also detects other classes of compounds (numerous examples are also shown in **Table 1**).

The routine use of IPAD for amino acid analysis is relatively new. However, many reports on IPAD of amino acids have been published during the past 15 yr *(5,7–12)*, and a comprehensive book *(10)* dedicated to the technique has recently become available.

Table 3
Group Gradient: Gradient Conditions for Simultaneous Separations of Amino Acids and Carbohydrates

Time (min)	%E1	%E2	%E3	Curve	Comments
Init	94	6	0		Autosampler fills the sample loop
0.00	94	6	0		Valve from Load to Inject
2	94	6	0		Begin OH gradient, valve from Inject to Load
12	68	32	0	8	
16	68	32	0		Begin acetate gradient
24	36	24	40	8	
40	36	24	40		
40.1	20	80	0	5	Column wash with hydroxide
42.1	20	80	0		
42.2	94	6	0	5	Start of reequilibration to initial conditions
62	94	6	0		

Table 4
Glucose Gradient: Gradient Conditions for Simultaneous Separations of Amino Acids and Glucose

Time (min)	%E1	%E2	%E3	Curve	Comments
Init	84	16	0		Autosampler fills the sample loop
0.00	84	16	0		Valve from Load to Inject
2	84	16	0		Begin OH gradient, valve from Inject to Load
12	68	32	0	8	
16	68	32	0		Begin acetate gradient
24	36	24	40	8	
40	36	24	40		
40.1	20	80	0	5	Column wash with hydroxide
42.1	20	80	0		
42.2	84	16	0	5	Start of reequilibration to initial conditions
62	84	16	0		

Like most other contemporary electrochemical techniques, IPAD is carried out in a three-electrode detection cell (**Fig. 4A**). The gold working electrode is subjected to several carefully chosen potential steps (E1 to E6 in **Fig. 4B**). The

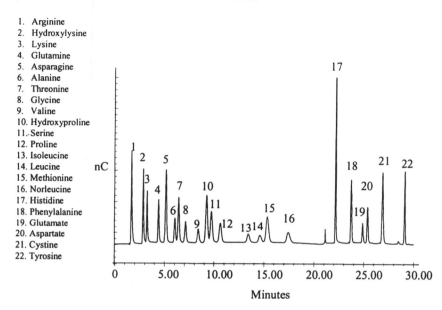

1. Arginine
2. Hydroxylysine
3. Lysine
4. Glutamine
5. Asparagine
6. Alanine
7. Threonine
8. Glycine
9. Valine
10. Hydroxyproline
11. Serine
12. Proline
13. Isoleucine
14. Leucine
15. Methionine
16. Norleucine
17. Histidine
18. Phenylalanine
19. Glutamate
20. Aspartate
21. Cystine
22. Tyrosine

Fig. 1. Separation of amino acids using the Standard Gradient Conditions in **Table 2**.

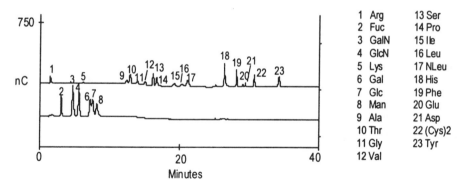

1 Arg	13 Ser
2 Fuc	14 Pro
3 GalN	15 Ile
4 GlcN	16 Leu
5 Lys	17 NLeu
6 Gal	18 His
7 Glc	19 Phe
8 Man	20 Glu
9 Ala	21 Asp
10 Thr	22 (Cys)2
11 Gly	23 Tyr
12 Val	

Fig. 2. Simultaneous separations of amino acids, amino sugars, and glucose using the gradient from **Table 4**.

potentials are adjusted between the gold surface and a silver/silver chloride reference electrode (more recently in reference to a glass/silver/silver chloride combination electrode *(5)*, *see* **Table 5**). Some chemistries that occur at the gold electrode surface are described in **Fig. 4B**. Additionally, oxidation and sometimes also adsorption and oxidative desorption are observed in the presence of the analytes in **Table 1**. Electrons transferred to the gold surface during

Jandik et al.

1. Arginine
2. Hydroxylysine
3. Lysine
4. Galactosamine
5. Glucosamine
6. Glucose
7. Alanine
8. Threonine
9. Glycine
10. Valine
11. Hydroxyproline
12. Serine
13. Proline
14. Isoleucine
15. Leucine
16. Methionine
17. Norleucine
18. Histidine
19. Phenylalanine
20. Glutamate
21. Aspartate
22. Cystine
23. Tyrosine

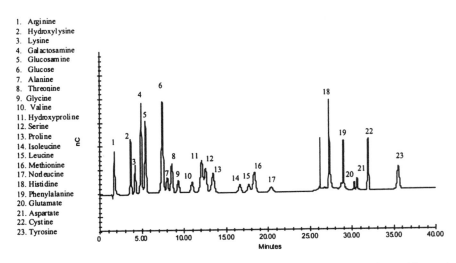

Fig. 3. Optimized gradient (*see* **Table 3**) allowing separations of amino acids, amino sugars, and sugars in one run.

analyte oxidation give rise to an electric current between the gold electrode and counterelectrode. The use of two separate electrodes (reference and counterelectrode) for voltage application and current sensing was introduced in the early days of electrochemistry. It helps to avoid significant electric currents running through the reference electrode, thereby keeping its potential stable over a long period. If the main portion of the current is allowed to pass through the reference electrode, it ââuses the silver chloride layer to dissolve from the silver wire and consequently modifies the value of the reference potential. The long-term stability of the reference potential is essential for good reproducibility of the detection response.

The current resulting from the analyte oxidation at the gold electrode is measured by integration over the time period indicated in **Fig. 4B**. The detection signal thus obtained is in nanocoulombs (coulomb = amperes second) rather than in amperes.

2. Materials

2.1. Equipment

1. Equipment for vacuum hydrolysis: Most common hydrolysis methods can be performed in borosilicate glass tubes equipped with Teflon plugs (P/N 29550, Pierce, Rockford, IL). Tubes are placed into aluminum heating block (P/N 18808, Pierce) and heated up by a Heating Module (P/N 18870 [120V] Pierce).
2. Equipment for evaporative centrifugation: Hermetically sealing centrifuge (SVC 100, Savant, Farmingdale, NY), refrigerated vapor trap (RVT 400, Savant), vacuum gage MVG-5 connected to DV-24 gauge tube (both from Savant), chemi-

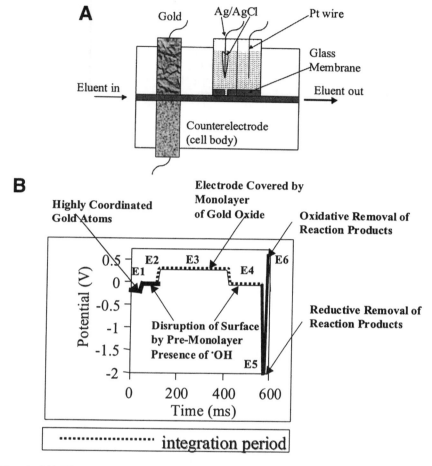

Fig. 4. (**A**) Three-electrode detection cell allowing a choice between silver/silver chloride and pH referencing. *See* **ref. 5** for more details. (**B**) The surface of the gold working electrode at different potentials of the detection waveform.

cal vapor trap (SCT 120, Savant) containing acid adsobent cartridge (DC120A) and a vacuum pump (e.g., P/N 7893-F10, Thomas Scientific, Swedesboro, NJ) are interconnected by thick-walled, half-inch ID, Tygon tubing (e.g., P/N 9561-E28, Thomas Scientific).

3. HPLC hardware: AAA Direct System (Dionex Corporation, Sunnyvale, CA) is optimized for anion exchange/IPAD of amino acids. It includes the AS50 PEEK autosampler, a Thermal Compartment, GP50 PEEK microbore pump with Vacuum Degas Option, and an ED40 electrochemical detector with AAA-Certified Gold Amperometry Cell. Mobile phase components are kept under helium or nitrogen at all times. For fully automatic operation, the PeakNet Control Windows 98 Workstation should also be selected.

Table 5
Detection Waveform for Amino Acids, Amino Sugars,
and Sugars

Time (ms)	Potential (V)[a] vs pH	Integration
0	0.13	
40	0.13	
50	0.28	
110	0.28	Begin
120	0.61	
410	0.61	
420	0.28	
560	0.28	End
570	−1.67	
580	−1.67	
590	0.93	
600	0.13	

[a]All potentials as in **Fig. 4B** +0.33V.

4. Anion-exchange column (*see* **Note 1**): Dionex distributes the high-pH-stable an-ion–exchange columns optimized for the separation of amino acids and other compounds under the name "AminoPac PA10" (*see* **Note 1**). They are typically used as a combination of guard column (2-mm id, 50-mm long) and main column (2-mm id, 250 mm long).

2.2. Reagents (see Note 2)

1. Oxidation and/or hydrolysis: formic acid (0128-01, J.T. Baker, Phillipsburg, NJ), 30% hydrogen peroxide (0128-01, J.T. Baker), 6 *N* HCl (24309, Pierce), 4.0 *M* MSA (25555, Pierce), 4.2 *M* NaOH (30,657-6, Sigma, St. Louis, MO), 13 *M* TFA (53102, Pierce).
2. Chromatographic eluents: 18 megohm water from a Milli-Q (Millipore, MA), 50% sodium hydroxide (UN1824, Certified Grade, Fisher, Pittsburgh, PA), anhy-drous sodium acetate (71179, Microselect Grade, Fluka, Milwaukee, MI).
3. Diluents and standards: norleucine (N1398, Sigma), sodium azide (S 8032, Sigma), Standard Reference Material Amino Acids (SRM 2389 NIST, Gaithersburg, MD), cysteic acid (C7630, Sigma), methionine sulfone (M0751, Sigma), methionine sulfoxide (M1001, Sigma), Mixture of Six Carbohydrates (P/N 43162, Dionex).

2.3. Preparation of Eluents and Standards (see Notes 3–6)

1. Eluent E1: Deionized water — Filter the pure deionized water through 0.2-μm nylon filters, then transfer into a DX-500 eluent bottle. Use, for example, steril-

ized, sterile-packed, 1-L-funnel, vacuum-filtration units from Nalge (28194–514, VWR, West Chester, PA). We have found these ideal for filtration of all eluents. Seal the filtered water immediately.

2. Eluent E2: 250 mM sodium hydroxide — The first step in preparing sodium hydroxide eluent is to filter a water aliquot (typically 1.0 L) using the sterilized Nalgene filtration unit as aforementioned. Hermetically seal the filtered water immediately after filtration while preparing a disposable glass pipet (10.0-mL sterile, serological pipets, Fisher) and a pipet filler. Using a pipet filler, draw an aliquot of 50% sodium hydroxide into the pipet. Most serological 10.0-mL pipets can be filled to the 13.1-mL volume required for 1.0 L of 250 mM sodium hydroxide. Unseal the filtered water and insert the full pipet approx 1 in. below the water surface, and release the sodium hydroxide. If done properly and without stirring, most of the concentrated sodium hydroxide remains at the lower half of the container and the carbon dioxide adsorption rate is much lower than with a 250 mM sodium hydroxide solution. Seal the container immediately after the sodium hydroxide transfer is complete. Remember to immediately put the screw-cap back on the 50% hydroxide bottle. Mix the contents of the tightly sealed container holding the 250 mM hydroxide. Unscrew the cap of the eluent bottle E2 attached to the DX-500. Allow the helium or nitrogen gas to blow out of the cap. Unseal the bottle holding 250 mM hydroxide, and immediately start the transfer into the eluent bottle E2. Try to minimize the carbon dioxide absorption by holding the gas orifice of the bottle cap as close as possible to the 250 mM hydroxide during the transfer. Follow aseptic procedures. With the inert gas still blowing, put the cap back on the eluent bottle. Allow the pressure to build inside the bottle and reopen the cap briefly several times to allow trapped air to be gradually replaced by the inert gas.

3. Eluent E3: 1.0 M sodium acetate — Dissolve 82.04 g of anhydrous sodium acetate in approx 750 mL of pure deionized water. Seal the container during the dissolution step. Make up to 1.0 L with water and filter through a 0.2-μm Nylon filter. We recommend using sterile Nalgene vacuum filtration units for this step (*see* above). Transfer the filtered sodium acetate eluent into the eluent bottle E3 of the DX-500, using the same procedure as for the sodium hydroxide.

4. Diluent solutions containing sodium azide: Prepare 1 or 2 L of an azide diluent solution containing 20 mg NaN_3/L. Use the solution without norleucine whenever an internal standard is not required. Store the pure azide solution in a closed container at room temperature.

5. Diluent solution containing norleucine internal standard: Prepare a 4.0 mM stock solution of norleucine (524.8 mg/L) in 0.1 M HCl. Dilute 500x with a deionized water solution containing 20 mg azide per liter. The resulting diluent solution is stable for months if stored in a refrigerator. Use it to prepare final dilutions from standard stock solution and to redissolve hydrolysate samples after evaporation to dryness.

6. Amino acid standards: Use aliquots of Standard Reference Material 2389 diluted 500x or 250x with the diluent (*see* **Subheading 2.3.**, **step 5**) to obtain 5 μM or 10

μ*M* standard solutions. The standard solutions thus prepared remain stable for months if stored in a refrigerator. The trace of sodium azide introduced with the diluent solution stabilizes standards up to 48 h at room temperature.

7. Standard mixture of oxidation products: Prepare a 5 m*M* stock solution of oxidation standards by dissolving 82.6 mg of methionine sulfoxide, 90.6 mg of methionine sulfone, and 84.6 mg of cysteic acid in 100.0 mL of 0.1 *M* hydrochloric acid. Transfer 100 μL of the stock solution to a 100 mL volumetric flask and fill up to volume with the diluent from **Subheading 2.3., step 5**.

8. Carbohydrate standard: Add 1.0 mL of sodium azide diluent (**Subheading 2.3., step 4**) to the Dionex carbohydrate standard to obtain a 100-μ*M* stock solution of each of the six components. Vortex to support the dissolution of the dry residue. Using the sodium azide diluent again, dilute an aliquot of the stock solution to obtain a 2.0 μ*M* concentration of all four sugars and two amino sugars.

3. Methods

3.1. Oxidation of Methionine and Cystine/Cysteine Prior to Hydrolysis

1. Prepare the performic acid reagent freshly from formic acid and hydrogen peroxide before each application. Mix 900 μL of the formic acid and 100 μL 30% hydrogen peroxide in a microcentrifuge tube. Vortex the mixture.
2. Allow the mixture to adjust to room temperature.
3. Pipet an aliquot containing about 1 μg of the sample into another microcentrifuge vial and remove the liquid by evaporative centrifugation.
4. Add 250 μL of the chilled performic acid solution to the dry sample. Seal the microcentrifuge vial containing the reaction mixture, and keep it at room temperature for 4 h.
5. Carry out evaporative centrifugation to remove the rest of the performic acid solution from the sample.
6. Proceed with hydrolysis (go to **Subheading 3.2., step 2**. Addition of 6 *M* HCl).

3.2. Hydrolysis (see Notes 7 and 8)

1. Pipet a volume of protein or peptide solution into a clean microcentrifuge tube. Carry out evaporative centrifugation to remove the liquid phase.
2. To a dry sample containing no more than 200 μg of protein or peptide, add 200 μL of the hydrolysis agent (6 *M* HCl, 2 *N* TFA, 4.0 *M* MSA, or 4.2 *M* NaOH). If you are processing a performic-acid-oxidated sample, add 200 μL of 6 *M* HCl to the dry sample from **Subheading 3.1., step 5**.
3. Speed up the dissolution of the solid residue in the hydrolysis agent by vortexing. Transfer the liquid contents into a vacuum hydrolysis tube (*see* **Note 9**).
4. Insert the Teflon plug, and screw it down leaving only a narrow passageway open. Attach the side arm of the tube to the vacuum tubing connected to a three-way stopcock. Connect the stopcock to a vacuum pump and to a source of inert gas.
5. Alternate between vacuum and inert gas to remove all traces of oxygen from the space above the hydrolysis mixture.

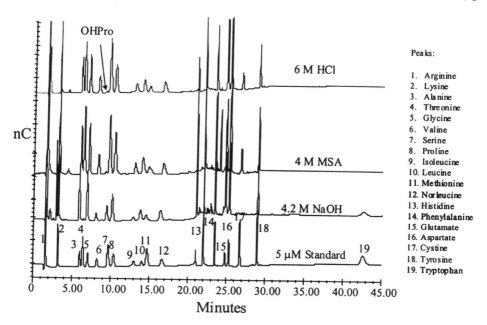

Fig. 5. Soybean powder: Chromatographic results from three different hydrolytic procedures.

6. Screw down the Teflon plug to seal the vacuum inside the tube.
7. Place the sealed tube into a preheated Reacti-Therm Module, and hydrolyze (*see* **Note 7** for temperature setting and timing of the respective hydrolytic protocols).
8. Remove the tube from the heating module and allow it to cool before breaking the vacuum inside the tube.
9. Transfer the hydrolysate into a clean microcentrifuge tube.
10. Carry out evaporative centrifugation (*see* **Note 10**).
11. Redissolve the dry contents of the microcentrifuge tube in a suitable volume of azide solution with/without internal standard (typically 1x to 4x of the volume during the hydrolysis). Dry samples from **step 10** are stable for up to 1 yr if stored in a freezer. The redissolved samples can be stored for up to 1 mo in a refrigerator. At room temperature, the azide-stabilized samples remain unchanged for at least 2 d. Proceed with chromatographic separation.

3.3. Chromatographic Separations of HCl, MSA, and NaOH Hydrolysates (see Note 11; Fig. 5)

The Standard Gradient in **Table 2** (*see* **Note 12**) offers a useful separation of hydrolysate amino acids and all other compounds listed in regular print in **Table 1**. The italics indicate the need for a modified gradient. All modified gradients, however, utilize the same three eluents (E1, E2, and E3; *see* **Subheading 1.1.**)

and are carried out at a constant flow rate of 0.250 mL/min. Selectivity changes are achieved by varying the initial gradient conditions or by changing the separation temperature (*see* **Notes 13, 14**).

1. Verify proper operation of the chromatographic system as described in the Dionex System Manual (background signal at initial gradient conditions < 80 nC, gradient rise < 30 nC, *see* Dionex Document No. 031481, **Notes 15–17**).
2. Transfer the samples from **Subheading 3.2.**, **step 11** into 1.5- or 0.5-mL autosampler vials (*see* **Notes 18–20**). Place vials into the autosampler, note their positions.
3. Fill a 1.5-mL vial with 8 μM solution of hydrolysate standard. Fill another 1.5-mL vial with water. Place vials into the autosampler, note their positions.
4. Write a PeakNet Schedule (*see* PeakNet Software User's Guide, Dionex Document No. 034914) specifying 25.0-µL full-loop injections and assigning the gradient pump method from **Table 2** together with the detection method from **Table 5** to each of the samples and standards. The sequence of injections should start with at least one injection of water, followed by one or two injections of the standard. Ideally, there should be at least two standard injections at the beginning, in the middle, and at the end of each sequence of injection.
5. Load the PeakNet Schedule from **Subheading 3.3.**, **step 4** into the PeakNet Run Window and start the automatic run.

3.4. Chromatographic Analysis of Oxidation Products of Cysteine and Methionine

1. Equilibrate the HPLC system using the initial conditions of standard gradient in **Table 2** with the column thermostat set at 35°C (*see* **Notes 13** and **21**). The background signal should be <100 nC.
2. Fill a 1.5-mL vial with water. Place vial into autosampler. Note its position.
3. Fill a 1.5-mL vial with 8 mM solution of hydrolysate standard. Place vial into autosampler. Note its position.
4. Fill a 1.5-mL vial with 5 µM solution of oxydation products standard. Place vial into autosampler. Note its position.
5. Transfer the samples from **Subheading 3.2.**, **step 11** into 0.5- or 1.5-mL autosampler vials (**Note 20**). Note their positions.
6. Write a PeakNet Schedule (*see* PeakNet Software User's Guide, Dionex Document No. 034914) specifying 25.0-µL full-loop injections and assigning the gradient method from **Table 2** together with the detection method from **Table 5** to each of the samples and standards (verify the 35°C column temperature setting). The sequence of injections should start with at least one injection of water, followed by one or two injections of each of the two standards. Ideally, there should be two injections of each of the two standards at the beginning, in the middle and at the end of each sequence of injection.
7. Load the PeakNet Schedule from **Subheading 3.4.**, **step 6** into the PeakNet Run Window, and start the automatic run (**Fig. 6**).

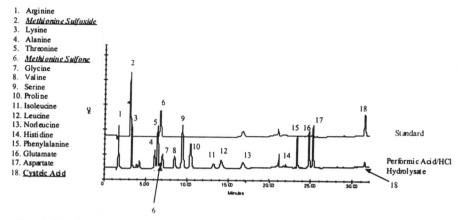

1. Arginine
2. *Methionine Sulfoxide*
3. Lysine
4. Alanine
5. Threonine
6. *Methionine Sulfone*
7. Glycine
8. Valine
9. Serine
10. Proline
11. Isoleucine
12. Leucine
13. Norleucine
14. Histidine
15. Phenylalanine
16. Glutamate
17. Aspartate
18. *Cysteic Acid*

Fig. 6. Bovine Colostrum Hydrolysate: Analysis of oxidation products of cystine/ cysteine and methionine. The separation of a mixture consisting of methionine-sulfoxide, methioninesulfone, and cysteic acid is included for easier identification of oxidation products. The methioninesulfoxide is not detected in the sample hydrolysed by HCl. To separate methioninesulfone from the neighboring threonine and glycine, the column temperature has to be maintained at 35°C. Since tyrosine is destroyed during the oxidative step, cysteic acid, the product of cysteine/cystine oxidation, elutes as an isolated peak at approx 32 min.

3.5. Chromatographic Analysis of Sugars, Amino Sugars, and Amino Acids in Hydrolysates Using a Single Gradient Method

The gradient method in **Table 3** makes possible a simultaneous separation of sugars, amino sugars and amino acids, *see* the chromatograms in **Fig. 7**.

1. Hydrolyze three aliquotes of a glycoprotein using the procedures in **Note 7A–C** and the hydrolytic protocol **Subheading 3.2., steps 1–11**.
2. Verify proper operation of the chromatographic system as described in the Dionex System Manual (background signal at initial gradient conditions < 80 nC, gradient rise < 30 nC, *see* Dionex Document No. 031481).
3. Fill a 1.5-mL vial with water. Place vial into autosampler. Note its position.
4. Fill a 1.5-mL vial with 8 m*M* solution of hydrolysate standard. Place vial into autosampler. Note its position.
5. Fill a 1.5-mL vial with 8 m*M* solution of hydrolysate standard (from **Subheading 2.3., step 6**). Place vial into autosampler. Note its position.
6. Fill a 1.5-mL vial with 2 m*M* solution of carbohydrate standard (from **Subheading 2.3., step 8**). Place vial into autosampler. Note its position.
7. Transfer the samples from step **Subheading 3.2., step 11** of each of the three hydrolytic protocols in **Note 7A–C** into 0.5-mL autosampler vials (**Note 20**).
8. Place sample vials into autosampler. Note their position.

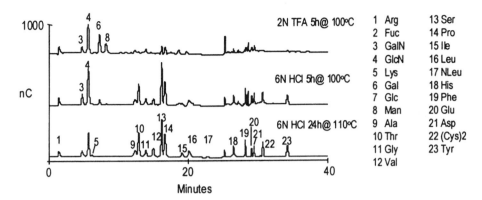

Fig. 7. Identical amounts (200 µg) of fetuin were hydrolyzed by three different procedures. The volumes of hydrolyzing agents were 200 µL 2 *N* TFA and 400 µL 6 *M* HCl. The hydrolysates were dried down by evaporative centrifugation and redissolved in 1.0 mL of diluent (**Subheading 2.3.5.**) All three hydrolysates could be analyzed by the same gradient method (**Table 3**).

9. Write a PeakNet Schedule (*see* PeakNet Software User's Guide, Dionex Document No. 034914) specifying 25.0-µL full-loop injections and assigning the gradient method from **Table 3** together with the detection method from **Table 5** to each of the samples and standards. The sequence of injections should start with at least one injection of water, followed by one or two injections of each of the two standards. Ideally, there should be two injections of each of the two standards at the beginning, in the middle, and at the end of each sequence of injection.
10. Load the PeakNet Schedule from **Subheading 3.5.**, **step 9** into the PeakNet Run Window and start the automatic run.

3.6. Data Processing Following Chromatographic Separation of Hydrolysate Samples (Notes 22–25)

1. At the conclusion of the sequence defined in step **Subheading 3.3.**, **step 5**, **Subheading 3.4.**, **step 7**, or **Subheading 3.5.**, **step 10**, inspect the standard chromatograms first. In a properly operated system, the reproducibility of peak areas is < 5% RSD and reproducibility of retention times is < 1% RSD. If your results do not conform with the %RSD values specified for the standard injections, optimize integration and repeat postrun data processing (*see* Verification of Reproducibility, Dionex LPN. 01156-01).
2. Create a Component Table listing amino acids (or amino acids, sugars, amino sugars and oxidation products, if applicable) of interest, retention times, and mean values of peak areas for the concentrations of the injected standard (*see* PeakNet Software User's Guide, Dionex Document No. 034914). Using the optimized peak integration from **Subheading 3.6.**, **step 1**, apply postrun processing to the hydrolysate samples to calculate amino acid concentrations (or sugar, amino

sugar, amino acid, and oxidation products concentrations, if applicable) in the hydrolysate samples.

3.7. Off-Line Analysis of Cell-Culture Samples (see Note 26)

The analysis of amino acids and sugars in cell cultures is part of a broader strategy aimed at maximizing the product titer *(9,16)*. Without analytical control, glucose concentration has to be maintained at excessively high levels, whereas the concentration of glutamine, another important nutrient, may decrease to the levels too low for the process to be efficient. The strong rise in acidity stemming from excessive lactic acid formation is neutralized by adding sodium hydroxide to the cell culture at regular intervals.

With the help of chromatographic analysis, an improved feeding strategy can be designed that makes it possible to replenish the glucose levels according to actual demand and keeps it at much lower levels. Lower levels of glucose result in less frequent additions of sodium hydroxide and a slower increase of ionic strength. The level of glutamine can also be kept relatively low and within very narrow boundaries. The production of ammonia is minimized and depletion of glutamine avoided.

1. Verify proper operation of the chromatographic system as described in the Dionex System Manual (background signal at initial gradient conditions < 80 nC, gradient rise < 30 nC, *see* Dionex Document No. 031481).
2. Equlibrate the HPLC system using gradient conditions in **Table 4** and the detection method from **Table 6**.
3. Run at least three 25-μL injections of standard (use NIST standard from **Subheading 2.3., step 6**) and complete the PeakNet Component Table (*see* PeakNet Software User's Guide, Dionex Document No. 034914).
4. Take samples (1–5 mL) of fermentation at timed intervals.
5. Using a sterilized syringe, pass an aliquot of the sample through a sterile 0.45-μm filter.
6. Dilute an aliquot of each filtered sample 250x using the diluent from **Subheading 2.3., step 4**.
7. Inject 25 μL of the dilute sample into the chromatographic system.
8. Initiate a run using a pump method based on **Table 4** gradient (*see* **Note 27**) and Table 6-detection method (*see* **Note 28**). Make the Component Table from **Subheading 3.6., step 2**, also a part of the PeakNet method, to obtain automatic calculation of amino acid concentrations in the fermentation broth (**Fig. 8**).

4. Notes

1. The AminoPac PA10 columns are packed with 9-μm-diameter microporous resin beads, consisting of ethylvinylbenzene crosslinked with 55% divinylbenzene. The MicroBead layer agglomerated on the surface of these beads consists of 200-nm VBC-based (vinylbenzylchloride) latex particles functionalized with a precise

Table 6
Modified for the Nonselective and Selective Detection

Time (ms)	Potential (V)[a] vs pH	Integration to detect Amino acids, sugars, and amino sugars	OH-selective integration
0	0.13		
40	0.13		
50	0.28		
110	0.28		Begin
210	0.28	Begin	End
220	0.61		
460	0.61		
470	0.28		
560	0.28	End	
570	−1.67		
580	−1.67		
590	0.93		
600	0.13		

[a]See the waveform graphs in **Fig. 10**.

mix of quaternary/tertiary ammonium groups. This mix of functionalized groups has a controlled ratio, resulting in excellent selectivity for the separation of amino acids. The small particle size of the anion–exchange layer exhibits adequate mass transfer characteristics enabling high chromatographic peak efficiency. The highly crosslinked core of the AminoPac PA10 packing makes it possible to use organic solvents to facilitate column cleanup.

2. Use chemicals exactly as specified by catalog number and supplier. Consult with Dionex should one or more of the specified reagents be not available in your geographical area.

3. Minimize any extraneous contamination of eluents. For example, a trace of an ion-pairing agent introduced into the eluent from a "shared" filtration apparatus will cause interference with some of the amino acid peaks. Dedicate glassware, pipets, filtration apparatus for exclusive use in preparation of AAA eluents only. Wear disposable powder-free gloves whenever preparing or refilling eluents.

4. Minimize the level of carbonate introduced into the eluents during preparation.

5. Avoid bacterial contamination of eluent bottles and tubing. The bacterial contamination is minimized by wearing disposable, powder-free gloves, keeping containers closed whenever possible, and by ultrafiltration (filter pore size < 0.2 μm). Use ultrafiltration as indicated in the instructions for preparing each of the three mobile phases. Microorganisms, if present in the system, produce amino acids, causing elevated background levels and spurious peaks.

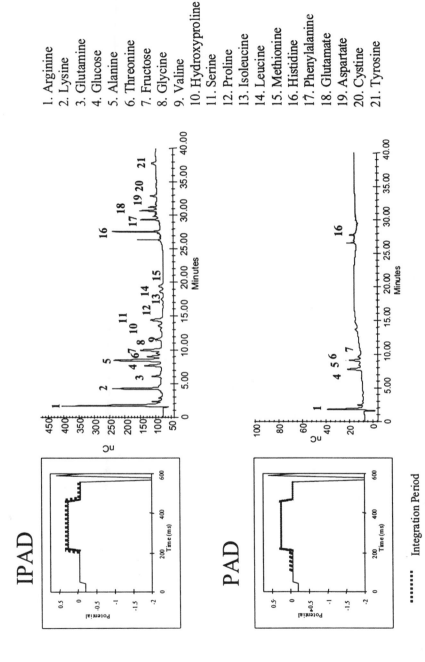

Fig. 8. Selective IPAD detection applied to a 1:250 diluted cell culture sample. The lower trace obtained with the hydroxy specific detection mode (*see* **Note 28**) helps to identify an unknown peak (Peak #7) as fructose. The identification of fructose in the upper chromatogram is difficult since it coelutes closely with glycine.

1. Arginine
2. Lysine
3. Glutamine
4. Glucose
5. Alanine
6. Threonine
7. Fructose
8. Glycine
9. Valine
10. Hydroxyproline
11. Serine
12. Proline
13. Isoleucine
14. Leucine
15. Methionine
16. Histidine
17. Phenylalanine
18. Glutamate
19. Aspartate
20. Cystine
21. Tyrosine

········ Integration Period

79

6. Remember, atmospheric carbon dioxide adsorbs even into pure water, albeit at much lower levels than in alkaline solutions. Minimize the time the surface of the water comes in contact with the atmosphere.

7. Follow the generally accepted hydrolysis procedures. The present method works with the protocols *(13–15)* listed below (**Fig. 5**):
   a. Monosaccharide analysis: 2 *M* TFA (31 μL of 13 *N* TFA plus 169 μL of water, 4–5 h, 100°C.
   b. Amino sugar analysis: 6 *M* HCl 4–5 h, 100°C.
   c. Amino acid analysis: 6 *M* HCl 18–24 h, 110°C.
   d. Amino acid analysis: 4.0 *M* MSA 24 h, 110°C.
   e. Tryptophan analysis: 4.2 *M* NaOH 24 h, 110°C.

8. The HCl hydrolysates usually show less interference compared to other protocols. Hydrolysis by MSA is preferred by many workers for analyzing plant related materials or animal tissue. Whenever using MSA hydrolysis for cellulose (plants) or carbohydrate (processed meats) containing materials, it is advisable to employ a gradient resolving glucose from alanine (**Subheading 3.5.**, **step 9**). Unlike the nynhydrin/cation approach, the IPAD technique is very compatible with NaOH hydrolysates. Neither evaporation nor neutralization is required. We have found that an accurate analysis of tryptophan is easier with NaOH hydrolysis, especially for plants and animal tissues (**Fig. 5**).

9. Always check blanks before reusing the hydrolysis tubes. Chemical cleaning (chromic acid, commercial cleaning formulations) does not always work. Even the most reliable procedure (heating the tubes to 500°C in a muffle furnace) may have to be repeated to obtain blanks free of carryover impurities.

10. HCl hydrolysates containing high enough concentrations of amino acids (0.1–2 μg/200 μL) do not require an evaporative centrifugation step. They can be injected after a 100- or 1000-fold dilution with the norleucine/azide diluent (**Subheading 2.3.**, **step 5**). All hydrolysates in sodium hydroxide can be injected directly. Neither a neutralization nor evaporation is required. All MSA hydrolysates should be diluted at least 1000-fold.

11. Sample pH adjustments are not required with the present chromatographic method.

12. The standard gradient runs and all AminoPac PA10 chromatographic runs can be subdivided into four parts:
   a. First, amino acids are eluted by the hydroxide eluent only. This is started by an isocratic composition and finished off by a gradient. In the **Fig. 1** chromatogram, this part is represented by a group of peaks between arginine and norleucine.
   b. To elute the second group of peaks (histidine through tryptophan) requires a more strongly eluting species, the acetate anion. The sharp spike (not numbered) between norleucine and histidine peak is always present and indicates the point where the conversion of the anion exchanger from hydroxide to acetate is complete. The spike itself consists of hydroxide ions and illustrates the focusing effect of the acetate anion on all the other ions still retained on the column at the end of the first part of the chromatographic run.

The second part of the run is complete when the acetate concentration has dropped to zero, and for 2 min the hydroxide concentration is increased to 200 mM.

   c. This third step accelerates the conversion of the anion exchange groups back to the hydroxide form, it also effects the removal of traces of carbonate brought on the column as an impurity in E1, E2, or E3.

   d. In the fourth step, the column is readjusted by pumping of initial gradient composition of sodium hydroxide.

13. The separation temperature must be controlled within ±1°C. The prescribed temperature is 30°C for most of the existing methods. The comparison of four chromatograms obtained between 30 and 45°C (**Fig. 9**) shows that hydroxyproline, serine, and methionine are most affected by temperature changes. Only a small effect is discernible on the amino sugars and threonine. Also, all the temperature-induced shifts are within the "hydroxide part" of the chromatogram. The position of the peaks elueted by acetate is virtually unchanged.

14. The position of serine relative to proline (**Fig. 9**) is a good indication of whether or not the calibration of the column thermostat is correct. At lower temperatures, that peak closely coelutes with proline and at a temperature that may be too high it moves closer to valine.

15. Do not polish the Dionex AAA Gold electrode. Polishing the gold electrodes with 1 μm alumina is a standard procedure in connection with the analysis of carbohydrates by PAD. Since IPAD is carried out at higher potentials than PAD, the microscopic roughness resulting from polishing leads to excessive background levels. Alternative procedures such as electropolishing and chemical cleaning are described in the Dionex AAA Direct System Manual.

16. Do not allow the pH and Ag/AgCl combination electrode to dry out. The electrode dries out easily and is irreversibly damaged if left in a disconnected detector cell or if stored dry. Follow the storage instructions in the Dionex ED40 manual.

17. The ED40 reference electrode should be checked against "known good" reference electrodes in regular intervals. A known good reference electrode is, for example, another unused reference electrode of the same type that was properly stored in a 3 M KCl solution. Discard any reference electrode if its potential differs by more than 50 mV from the known good reference. Use only a high impedance voltmeter (e.g., a laboratory pH meter) to check reference potentials.

18. Sample tray cooling is available with the Dionex AS50 autosamplers.

19. The azide-containing diluent solution stabilizes samples and standards for at least three days. Overnight and over-the-weekend runs are possible even without a sample tray cooling.

20. Use low-volume autosampler vials (V = 300 μL, PN 500114, Sun International, Wilmington, DE) for limited volumes of undiluted samples (< 0.5 mL). Use standard autosampler vials made of glass (V = 1.5 mL, Dionex 055427) for larger volumes (>0.5 mL).

1. Arginine
2. Hydroxylysine
3. Lysine
4. Galactosamine
5. Glucosamine
6. Alanine
7. Threonine
8. Glycine
9. Valine
10. Hydroxyproline
11. Serine
12. Proline
13. Isoleucine
14. Leucine
15. Methionine.
16. Histidine
17. Phenylalanine
18. Glutamate
19. Aspartate
20. Cystine
21. Tyrosine

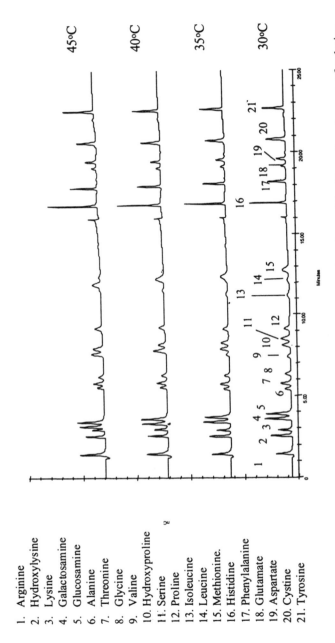

Fig. 9. Influence of column temperature on the separation of amino acids. Note the changes of relative positions of methionine vs leucine and hydroxyproline and serine.

82

21. The column temperature of 35°C has been found useful for separating amino acids in the presence of methionine sulfone and cysteic acid.

22. Evaluation of linearity in the range of 1 to 1000 pmol *(5)* showed the following results: Calibration plots are linear across three orders of magnitude for Ala, Thr, Gly, Val, Ser, Pro, Ile, Leu, Met, Phe, Glu, Asp, Cys, and Tyr. Calibration plots are linear over 2.25 orders of magnitude for Arg, Lys, Asn, Gln, and His.

23. The following limits of detection (3x noise in picomoles) were determined by injecting 10 pmol of each compound under the conditions of **Fig. 2**: Arg 0.15, Lys 0.21, Gln 0.48, Asn 0.35, Ala 0.72, Thr 0.40, Gly 0.62, Val 1.48, Ser 0.49, Pro 1.29, Ile 2.84, Leu 4.42, Met 1.12, His 0.11, Phe 0.36, Glu 0.78, Asp 0.51 (Cys) 2 0.26, Tyr 0.44.

24. If a mixture of cysteine and cystine is injected, only a single peak is observed. Single component standards of cysteine or cystine each yield a peak at the identical retention time.

25. Analytical results obtained by the IPAD method are in a good agreement with the nynhydrin/cation–exchange technique for a large number of protein and peptides. Two reports comparing the results for ribonuclease A and collagen are available in the literature *(5,12)*.

26. The possibility of monitoring amino acids and glucose with an On-Line Chromatographic Analyzer (Dionex DX-800) has been demonstrated. Fully automatic sampling is achieved using a tangential membrane filtration device as a sterile barrier. All connecting tubing between the fermentor and the filter and that leading from the filter to the On-Line analyzer has to be carefully sterilized prior to use.

27. Glucose and amino acids can be quantified reliably by using the gradient method in **Table 4**. We have found it possible to achieve an automatic peak detection of alanine by the PeakNet software up to the glucose/alanine molar ratio of 200 to 1. Should the sample contain an even higher excess of glucose and/or additional carbohydrate components, it may become necessary to use the Group Separation Gradient from **Table 3**. Using that gradient, we have verified experimentally that automatic peak detection and quantitation of alanine are possible up to a 1000-fold molar excess of glucose.

28. IPAD detection of amino acids and carbohydrates relies on the presence of suitable catalysts on the gold electrode surface. While the detection of hydroxide groups in carbohydrates requires the presence of hydroxide radicals, a monolayer of gold oxide is necessary to make the oxidation of amino groups of amino acids possible. In most applications, both types of catalytic surfaces are sequentially generated during each cycle of the IPAD detection waveform (*see* **Fig. 4B**). A typical IPAD method detects amino acids by integrating over the entire length of step E3. The sugars are detected by adding the entire length of step E4 to the period of current integration. The codetection of carbohydrates is, to a degree, a side benefit; the main purpose of the "dual-step" integration is to eliminate any contribution of gold oxide formation to the detection signal. Oxidation and reduction currents cancel each other out during the E3 and E4 integration *(11)*.

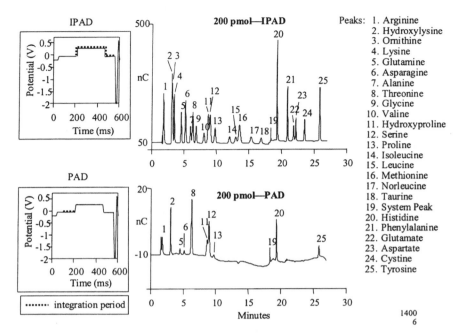

Fig. 10. Changes of selectivity in standard chromatograms with two different wave-forms. **Note:** the coeluting hydroxylysine is resolved from ornithine by the "hydroxy-selective waveform" in the lower chromatogram.

If the detection waveform is modified to expand the length of step E2 as shown in **Fig. 10**, it is possible to design a current-integration method that is sensitive for carbohydrates only. This relatively minor modification of the detection conditions enables us to choose between the simultaneous detection of amino acids and carbohydrates (**Fig. 10**, upper trace) or the more selective detection of only those molecules carrying hydroxy functional groups (**Fig. 10**, lower trace). The latter include not only sugar alcohols, amino sugars, and mono- and disaccharides, but also hydroxyamino acids.

The hydroxy-specific current-integration also detects the basic amino acids (arginine, lysine, glutamine, asparagine, and histidine, **Fig. 10**, lower trace) if they are present at high enough concentrations.

References

1. Kissinger, P. T., Refshauge, C., Dreiling, R., and Adams, R. N. (1973) An electro-chemical detector for liquid chromatography with picogram sensitivity. *Anal. Lett.* **6,** 465

2. Hughes, S. and Johnson, D. C. (1983) Triple-pulse amperometric detection of carbohydrates after chromatographic separation. *Anal. Chim. Acta* **149,** 1–10.

3. Edwards, P. and Haak, K. K. (1983) A pulsed amperometric detector for ion chromatography. *Am. Lab.* **April,** 78–87.
4. Polta, J. A. and Johnson, D. C. (1983) The direct electrochemical detection of amino acids at a platinum electrode in an alkaline chromatographic effluent. *J. Liq. Chromatogr.* **6,** 1727–1743.
5. Clarke, A. P., Jandik, P., Rocklin, R. D., Liu, Y., and Avdalovic, N. (1999) An integrated amperometry waveform for the direct, sensitive detection of amino acids and amino sugars following anion-exchange chromatography. *Anal. Chem.* **71,** 2774–2781.
6. Moore, S. and Stein, W. H (1963) Chromatographic determination of amino acids by the use of automated recording equipment, in *Methods in Enzymology*, vol. 6 (Colowick, S. P., and Kaplan, N., eds.), Academic, San Diego, CA, pp. 819–831.
7. Johnson, D. C. and LaCourse, W. R. (1990) Liquid chromatography with pulsed electrochemical detection. *Anal. Chem.* **62,** 589A–597A.
8. Johnson, D. C., Dobberpuhl, D., Roberts, R., and Vandeberg, P. (1993) Pulsed amperometric detection of carbohydrates, amines and sulfur species in ion chromatography — the current state of research. *J. Chromatog.* **640,** 79–96.
9. Jandik, P., Clarke, A. P., Avdalovic, N., Andersen, D. C., and Cacia, J. Analyzing mixtures of amino acids and carbohydrates using bi-modal integrated amperometric detection. *J. Chromatog.* **733,** 193–201.
10. LaCourse, W. R. (1997) *Pulsed Electrochemical Detection in High-Performance Liquid Chromatography*. Wiley – Interscience, New York.
11. Neuburger, G. G. and Johnson, D. C. (1988) Pulsed coulometric detection with automatic rejection of background signal in surface-oxide-catalyzed anodic detections at gold electrodes in flow-through cells. *Anal. Chem.* **60,** 2288–2293.
12. Martens, D. A. and Frankenberger, W. T. (1992) Pulsed amperometric detection of amino acids separated by anion exchange chromatography. *J. Liq. Chromatogr.* **15,** 423–439.
13. Davidson, I. (1997) Hydrolysis of samples for amino acid analysis, in *Methods in Molecular Biology*, vol. 64: Protein Sequencing Protocols (Smith, B. J., ed.), Humana, Totowa, NJ, pp. 119–129.
14. Smith, A. J. (1997) Postcolumn amino acid analysis, in *Methods in Molecular Biology*, vol. 64: *Protein Sequencing Protocols* (Smith, B. J., ed.), Humana, Totowa, NJ, pp. 139–146.
15. Web Page: *Amino Acid Analysis*, http://cmgm. stanford. edu/pan/aaasop.
16. Xie, L. and Wang, D. I. C. (1996) High cell density and high monoclonal antibody production through medium design and rational control in a bioreactor. *Biotechnol. Bioeng.* **51,** 725–729.

8

Ion-Pair Chromatography for Identification of Picomolar-Order Protein on a PVDF Membrane

Noriko Shindo, Tsutomu Fujimura, Saiko Kazuno, and Kimie Murayama

1. Introduction

There are several highly sensitive methods for identifying proteins: N-terminal sequencing, peptide mapping, and Western blotting, which are based on partial information about the structure of the protein. The amino acid composition of a protein is reflected in the primary structure of the whole protein, and this information is highly conserved.

Recently, we have reported a highly sensitive method of amino acid analysis — 6-aminoquinolyl-carbamyl (AQC)-amino acid analysis — using ion-pair chromatography (*1*) and used it to identify multidrug resistant protein 1 (MDR1) of mouse leukemia P388 cells on a polyvinylidene fluoride (PVDF) membrane with the aid of World Wide Web (WWW)-accessible tools (*2*).

MDR1 could not be confirmed by immunostaining with a commercially available murine monoclonal antibody (MAbC219) because of low specificity for this protein. Membrane proteins from the cells were separated by 1-D sodium dodecyl sulfate-polyacrylamide gel electrophoresis (SDS-PAGE), and blotted onto a polyvinylidene difluoride (PVDF) membrane. The 160-kDa protein band was excised and hydrolyzed in a vapor phase, then subjected to *in situ* AQC-derivatization without extraction of the amino acids. The amino acid composition of the protein was obtained using a highly sensitive AQC-amino acid analyzer employing ion-pair chromatography, and the protein was identified on the basis of its amino acid composition using the AACompIdent tool in ExPASy and PROPSEARCH in EMBL via the WWW. The methods employed are described in this chapter.

From: *Methods in Molecular Biology*, vol. 159: *Amino Acid Analysis Protocols*
Edited by: C. Cooper, N. Packer, and K. Williams © Humana Press Inc., Totowa, NJ

The most important consideration for such highly sensitive analysis is to minimize or decrease the extent of contamination from the environment, and sample loss. This was done by sample manipulation in a hand-made clean hood with gloves as far as possible, and *in situ* AQC-derivatization without extraction of amino acids from the PVDF membrane.

2. Materials

2.1. Reagents

Table 1 shows the reagents and consumables employed.

2.2. Apparatus

1. Clean hood: This was made from a metal frame ($60 \times 60 \times 60$ cm), the top, back, and sides of which were covered with clean transparent vinyl sheets and taped. The front sheet was hung to allow manipulation, and on its outer surface, a glass plate (10×20 cm) was attached with a glued tape to facilitate a clear view (*see* Note 3). Before and after the experiment, the inside of the hood was sterilized with ethanol for disinfection (*see* Note 4).
2. Hydrolysis vessel: JASCO Co. (http://www.jasco.co.jp/) code no. R154 (*see* Fig. 1). Another commercially available hydrolysis vessel is described by Wilkins et al. *(3)*.
3. Shimadzu high-performance liquid chromatograph LC-10A system: Column oven CTO-10AC; Auto sample injector SIL–10AXL with sample cooler; Fluorescence detector RF-10AXL; LC-Workstation Class-LC10.
4. Other apparatus and tools are shown in Table 2.

3. Methods

Fig. 2 shows a flow chart of the manipulation under the clean hood, gloves, and arm covers.

3.1. SDS-PAGE and PVDF Membrane Blotting

1. Prepare a plasma-rich membrane fraction of leukemia cells (1×10^8 cells) by sonication (2×10 s), followed by centrifugation ($100,000g$ for 1 h at 4°C), and suspend in 1 mL of Dulbecco's phosphate-buffered saline (PBS) by ultrasonication (2×10 s).
2. Determine the protein concentration using BCA protein assay reagent (product no. 23225, Pierce, Rockford, IL).
3. Dissolve one aliquot (100 μg protein) in an equal volume of sample buffer, and load onto a 1D SDS-PAGE gel.
4. Use 10 μg of bovine serum albumin (BSA) as a calibration protein for determining the composition of the amino acids in this experiment (*see* Note 5).
5. Electrophorese at 5 mA for 15 h at room temperature.
6. Perform the transfer onto the PVDF membrane using the tank blotting method and 10 mM CAPS buffer (*see* Notes 6 and 7). Wash the membrane with MeOH

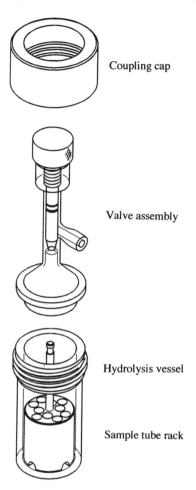

Coupling cap

Valve assembly

Hydrolysis vessel

Sample tube rack

Fig. 1. The design of the hydrolysis vessel. It is composed of a glass container, a valve assembly, a coupling cap, and a Teflon sample tube rack.

before use, and then soak in the transfer buffer for a few minutes. Equilibrate the gel in the buffer for 5 min.

7. Carry out the electrotransfer at 4°C for 1.5 h at 50 V.
8. Stain the membrane with 0.1% Coomassie brilliant blue R–350 for 5 min.
9. Destain quickly five times in 50% MeOH (each 30 s) in order to prevent protein loss, and the dry membrane in a refrigerator.
10. Determine the molecular weight (MW) using the image analyzer.

3.2. Acid Hydrolysis

1. Cut out the protein band on the PVDF membrane with a knife or a surgical blade, and cut this into 2 or 3 pieces, if necessary (*see* **Note 8**). To prevent the band from spreading, do not use a wet blade. Also, cut out a piece of blank membrane from

Table 1
Reagents

Method	Solution	Specification	Maker	Product no.
General Use	fresh Milli-Q Water	Analytical grade	PE Biosystems, CA, USA	015-08633
	Acetonitrile (MeCN)	HPLC grade	Wako Pure Chemical Industries Ltd., Osaka, Japan	
	Methanol (MeOH)	HPLC grade	Wako Pure Chemical Industries	052-03343
	Ethanol for disinfection	76.9–81.4%	Wako Pure Chemical Industries	328-00035
SDS-PAGE				
1D-PAGE	4% for stacking gel, 6.2% for separation gel	$13.5 \times 15 \times 0.1$ cm		
Buffer	125 mM Tris-HCl, pH 6.8			
Sample buffer	The above buffer containing 4% SDS, 10% 2-mercaptoethanol, and 0.0025% BPB			
Electroblotting				
Transfer buffer	10 mM CAPS buffer (3-[Cyclohexylamino]-1-propanesulfonic acid) in 10% MeOH, pH 11.0		Dojindo Laboratories, Kumamoto, Japan	343-00484
Staining solution	0.1% CBB (Coomassie brilliant blue [CBB] R-350) in 30% MeOH, 10% acetic acid	PhastGel BlueR	Amersham Pharmacia Biotech, Uppsala, Sweden	17-0518-01
Destaining solution	50% MeOH			17-0518-01
Hydrolysis				
Acid solution	6 N HCl containing 2% thioglycolic acid or 2.5% phenol	Amino acid analysis grade	Wako Pure Chemical Industries Ltd.	086-03925
	Hydrochloric acid (HCl)	for Nucleic acid extraction	Wako Pure Chemical Industries Ltd.	168-1272
	Phenol, crystal	Special grade	Wako Pure Chemical Industries Ltd.	204-01023
	Thioglycolic acid			

AQC-Derivatization

AHC Reagent	AccQ-Fluor™ Reagent Kit	Waters Co., Millford, MA	52880
	10 mM AHC [6-Aminoquinolyl-N-hydroxysuccinimidyl carbamate] in MeCN (**Note 1**)		
Solution	for Electrophoresis	Nacalai Tesque, Kyoto, Japan	052-18
	0.2 M Borate buffer pH8.8 (**Note 1**)		

Ion-Pair Chromatography

Column	4.6 mm ID × 150 mm	Nacalai Tesque, Kyoto, Japan	381-55
	Cosmosil 5C8-MS		
Elution buffer A	30 mM Phosphate buffer pH 7.2 containing 5 mM TBA-Cl		
	for ion-pair chromatography	Fluka Chemie AG, Buchs, Switzerland	86852
	Tetrabutylammonium chloride (TBA-Cl)		
Elution buffer B	The solution pH 7.2, which are mixed 260 mL of MeCN, 65 mL of 30 mM phosphate buffer and 175 mL of 30 mM KH_2PO_4 (**Note 2**)		
Amino acid solution	type H (2.5 μmol/mL, each)	Wako Pure Chemical Industries Ltd.	019-08393
	Wako amino acid standard solution		

Table 2
Apparatus and Tools

Name	Specification	Maker	Product no.
General Use			
Flush Mixer			13791
Oven	set at 300°C for pyrolyzation of glass wares	Sorenson™ BioScience, Inc., UT	
Microcapillary tips	83-mm length (0.5–200 μL)	Microflex Medical Corp., Malaysia	EV2050
Gloves	powder free latex gloves, disposable	Iuchiseieido, Osaka, Japan	88-1073-01
Disposable arm cover			
Electroblotting			
Minitransblotting apparatus	or equivalent apparatus	Nippon Eido, Co., Ltd., Tokyo, Japan	NC10115
Polyvinylidene fluoride (PVDF) membrane	ProBlott	PE Biosystems, CA	
Image Analyzer	analysis program: RFLP	Scanalytics, MA	
Hydrolysis			
Hydrolysis vessel	(see **Fig. 1**)	JASCO Co., Hachioji, Japan	R154
Hydrolysis tube and silicon plug or an inner glass tube	5.2 mm ID × 46.2 mm	JASCO Co., Hachioji, Japan	R158, R159
Aluminum block	set at 110°C.	JASCO Co., Hachioji, Japan	R157
Hydrolysis furnace			
Glass plate	for excising bands. 10 cm × 10 cm.		
Dry seal vacuum desiccator	grease-free	Wheaton, Millville, NJ	365884
0.45 μm Filter cartridge	Ekicrodisc 13	Gelman Science Japan, Tokyo, Japan	E131
Vaccum pump 1	Pumping speed 6L/min (suction of solution, with a glass trap) (see **Fig. 2**)	Sinku Kiko, Yokohama, Japan	G-5
Glass trap			
Vaccum pump 2	Pumping speed more than 100 L/min equipped with a solid-NaOH trap	Sato Vacuum Machinery Ind. Co., Ltd. Japan	DW-180
Surgical blade or knife	for excising protein band.	Feather Safety Razor Co., Ltd., Osaka, Japan	
AQC-Derivatization			
Ultrasonic bath	add fresh Milli-Q when use.	Branson Ultrasonic Co., CT	B-12
Heating bath	set at 55°C.	Riko-Kagaku Sangyo Co., Ltd., Japan	MH-10D

an area on which sample proteins have not run on. Place the standard, sample, and blank membrane fragments in separate glass tubes.

2. Wash the membrane pieces with 200 μL of 50% MeCN (or MeOH) on a mixer. Immediately suck up the washing solution with an aspirator, e.g., a handmade aspirator using pump 1. Excessive exposure to MeCN leads to sample loss (*see* **Note 9**).

3. While the membranes are wet, wash them with 200 μL of 0.01 *N* HCl, and then aspirate the solution. This washing step is done 2 or 3 times, and finally the solution takes place in 50% MeCN.

4. Dry the membranes in a vacuum desiccator for 10 min, and then slowly introduce the air into the desiccator through a 0.45-μm filter cartridge.

5. Add 6 *N* HCl (containing 2.5% phenol or 2% thioglycolic acid) to the hydrolysis vessels.

6. Place the tube in a Teflon sample tube rack, which is then placed into the vessel. Tightly close the cap and the valve of the vessel.

7. Move the vessel out of the hood, evacuate for 2 min, and place in a preheated aluminum block (110°C).

8. Carry out the hydrolysis at 110°C overnight.

9. Connect the valve of the hot vessel from the heating block to pump 2, and evacuate the residual 6 *N* HCl (roughly 5 min).

3.3. In Situ *Derivatization of AQC-Amino Acids*

1. Break vacuum of the vessel through a 0.45-μm filter cartridge.

2. Put the sample tube rack in a vacuum desiccator, and the membrane are completely dried with pump 1 for 10 min (*see* **Note 10**).

3. Moisten the membranes by dropping 10 μL of neat MeCN directly onto them, followed by 10 μL of 0.01 *N* HCl, and 10 μL of fresh Milli-Q water. AQC-derivatization of the standard amino acid mixture is done by adding 10 μL of standard solution in 0.01 *N* HCl instead of 0.01 *N* HCl for the membrane sample.

4. Add 50 μL of 0.2 *M* borate buffer and 20 μL of 10 m*M* AHC reagent, and leave the solution to stand for 1 min. Remove the sample rack from the hood and heat at 55°C for 10 min (*see* **Note 11**).

5. Transfer the derivatized solution to an autosampler plastic tube before HPLC analysis to avoid adsorption of amino acids onto the membrane and the wall of the glass tube.

6. Dilute 10 μL of the solution 10-fold with 0.2 *M* borate buffer.

3.4. AQC-Amino Acid Analysis

1. AQC-amino acids are separated by ion-pair chromatography using TBA-Cl.

2. The elution program is shown in **Table 3** (*see* **Notes 12** and **13**). The column temperature is kept at 30°C.

3. The elution profile of standard AQC-amino acids (1 pmol) is shown in **Fig. 3**.

4. Prepare the standard AQC-amino acid solution to derivatize 10 μL of 10 pmol/μL amino acid standard mixture, and store at –40°C (*see* **Note 14**). Dilute the AQC-

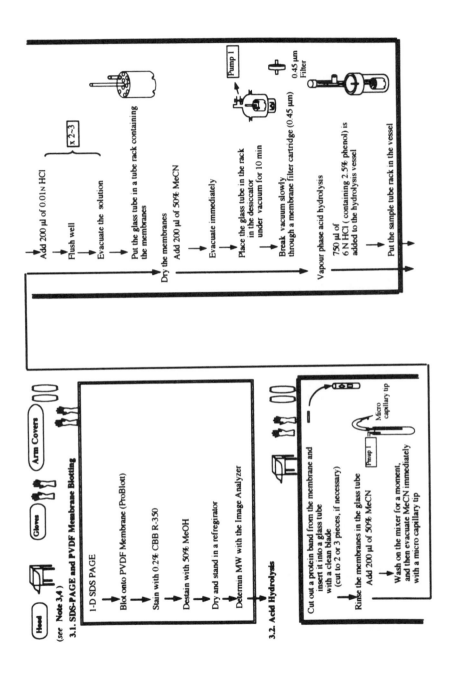

Heel **Gloves** **Arm Covers**

(see Note 3,4)

3.1. SDS-PAGE and PVDF Membrane Blotting

1-D SDS PAGE

→ Blot onto PVDF Membrane (ProBlott)

→ Stain with 0.2% CBB R-350

→ Destain with 50% MeOH

→ Dry and stand in a refregirator

→ Determin MW with the Image Analyzer

3.2. Acid Hydrolysis

Cut out a protein band from the membrane and insert it into a glass tube with a clean blade (cut to 2 or 3 pieces, if necessary)

→ Rinse the membranes in the glass tube

Add 200 µl of 50% MeCN

→ Wash on the mixer for a moment, and then evacuate MeCN immediately with a micro capillary tip

Micro capillary tip

Pmap 1

→ Add 200 µl of 0.01N HCl

→ Flush well

→ Evacuate the solution

x 2~3

→ Put the glass tube in a tube rack containing the membranes

Dry the membranes

Add 200 µl of 50% MeCN

→ Evacuate immediately

→ Place the glass tube in the rack in the desiccator under vacuum for 10 min

→ Break vacuum slowly through a membrane filter cartridge (0.45 µm)

Vapour phase acid hydrolysis

→ 750 µl of 6 N HCl (containing 2.5% phenol) is added to the hydrolysis vessel

→ Put the sample tube rack in the vessel

Pump 1

0.45 µm Filter

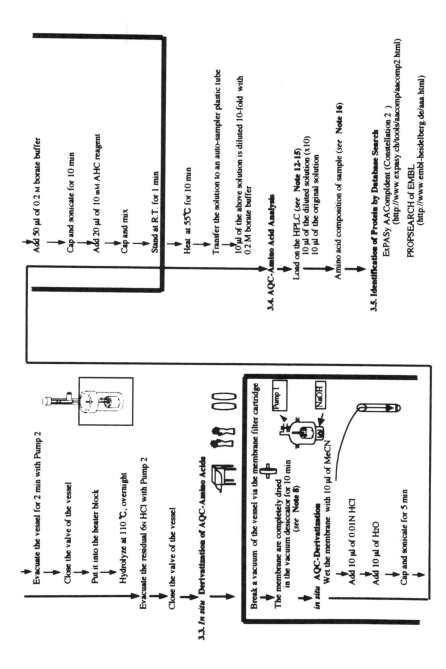

Fig. 2. Flowchart of manipulation under the hood with gloves and arm covers.

Table 3
Elution Program

Time (min)	Concentration of buffer B (%)
0.00	11.5
3.00	11.5
3.01	16.3
63.00	62.5
63.01	100.0
71.00	100.0
71.01	11.5
91.00	11.5

Buffer A: 30 mM phosphate buffer (pH 7.2), 5 mM TBA
Buffer B: 52% MeCN, 48% of 30 mM phosphate buffer
Flow rate: 0.5 mL/min
Column Temperature: 30C, isothermal
Detection: Ex 250 nm, Em 395 nm

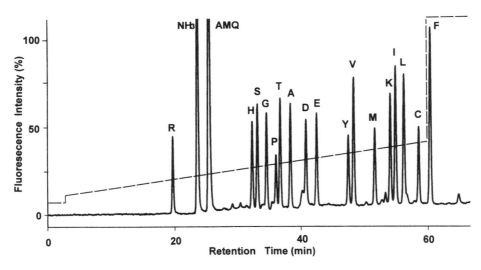

Fig. 3. Chromatogram of 1 pmol of the AQC-amino acid standard. AQC- amino acids are shown as one-letter code.

standard solution and sample 10-fold with 0.2 M borate buffer (pH 8.8), and inject 10-µL aliquots into the column (*see* **Note 15**).

5. **Figure 4** shows the 1-D SDS PAGE pattern for P388/S and P388/DOX cells **(A)** and the AQC-amino acid chromatogram for the 160-kDa protein band of P388/ DOX cells, showing drug resistance for doxorubicin **(B)**.

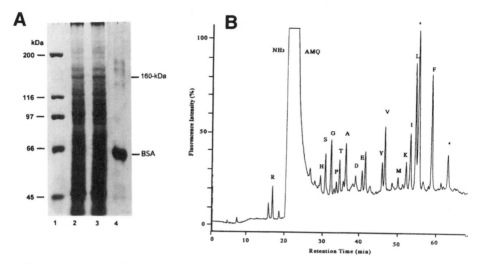

Fig. 4. (**A**) 1-D SDS-PAGE of P388/S and P388/DOX cells. Lane 1: Molecular weight standards, Lane 2: P388/S (control), Lane 3: P388/DOX, and Lane 4: BSA as a calibration standard. (**B**) AQC-amino acid chromatogram for the 160-kDa protein band of P388/DOX cells. *: Reagent-related peak.

6. Confirm the peak identification and peak area of each amino acid.
7. Calculate the concentration of the 16 amino acids against the standard amino acid mixture, and then export the data to a Microsoft Excel program.
8. Calculate the amino acid composition as the picomolar composition (%) in table form for the subsequent database search.
9. **Table 4** was prepared for the database search using ExPASy. It contained the amount of amino acid (pmol) and the amino acid compositions of P388/S, P388/DOX, and BSA that was used to correct the errors resulting from the hydrolysis (*see* **Note 16**).
10. Monitor the contamination of the amino acids using the membrane blank chromatogram.

3.5. Identification of Protein Using WWW Tools

1. Protein identification is done using its amino acid composition and MW and/or pI. These values are obtained from image analysis of the CBB-stained membrane. Protein identification is done using AACompIdent of the ExPASy molecular biology server (Constellation 2, for 16 amino acids) (http://www.expasy.ch/tools/aacomp/aacomp2.html) or PROPSEARCH of EMBL (http://www.embl-heidelberg.de/aaa.html).
2. The database search with PROPSEARCH is done using the composition corrected by the recovery of calibrated protein. The MW was determined to be 160-kDa from the blotted 1-D SDS-PAGE membrane using an image analyzer.

Table 4
Amino Acid Compositions of 160-kDa Protein in P388/S and P388/DOX Cells

Amino Acid	Standard (BSA) Experimental Amount (pmol)	Standard (BSA) Experimental Composition (pmol%)	P388/S Experimental Amount (pmol)	P388/S Experimental Composition (pmol%)	P388/DOX Experimental Amount (pmol)	P388/DOX Experimental Composition (pmol%)
Arg	4.22	5.16	0.27	9.04	0.17	5.37%
His	2.66	3.26	0.05	1.68	0.06	1.91%
Ser	4.26	5.21	0.21	7.10	0.22	7.03%
Gly	3.36	4.11	0.27	8.90	0.30	9.71%
Pro	3.74	4.57	0.09	3.10	0.09	2.90%
Thr	5.05	6.18	0.17	5.70	0.14	4.69%
Ala	6.91	8.45	0.20	6.66	0.19	6.11%
Asp	4.37	5.35	0.07	2.51	0.11	3.50%
Glu	6.79	8.31	0.18	6.10	0.18	5.84%
Tyr	5.13	6.27	0.17	5.65	0.17	5.65%
Val	6.78	8.29	0.26	8.61	0.30	9.85%
Met	0.60	0.74	0.08	2.83	0.06	1.88%
Lys	5.31	6.50	0.11	3.58	0.14	4.46%
Ile	2.01	2.46	0.16	5.49	0.24	7.72%
Leu	12.93	15.82	0.42	14.07	0.44	14.35%
Phe	7.63	9.33	0.27	8.99	0.28	9.02%
Total	81.74	100.00	2.98	100.00	3.08	100.00%

3. The results from database search with ExPASy are shown in **Fig. 5**. This shows the expression of MDR 1 in P388/DOX cells (score 14).

4. Notes

1. AccQ-Fluor™ Diluent and Borate Buffer in the kit were not used.
2. Elution buffer B in the original method (80% MeCN) was modified, and the elution program was changed to obtain the same separation as before, because of column maintenance.
3. A similar apparatus is commercially available. Our first clean hood used a computer rack as a frame.
4. It is important to prevent outside air from entering the hood. Therefore, it may be useful to spray the outside of the hood with ethanol for disinfection before use.
5. Usually, one of the SDS-PAGE MW standard bands is used as a calibration protein, but some of the prestained MW standards are not useful because they are largely modified from the original proteins.
6. The sponge pad, two filter papers, PVDF membrane, gel, two filter papers, and the sponge pad are wetted in the 10 m*M* CAPS buffer and place one after another

A P388/S

The SWISS-PROT entries having pI and Mw values in the specified range
for the species MOUSE and the specified keyword:

Rank Score Protein (pI   Mw) Description
==

1 34 PHYD_ARATH 5.81 129302 PHYTOCHROME D.
2 34 PHYB_ARATH 5.62 129331 PHYTOCHROME B.
3 35 CYA9_MOUSE 6.85 150954 ADENYLATE CYCLASE, TYPE IX (EC
4 35 KPB1_MOUSE 5.55 138793 PHOSPHORYLASE B KINASE ALPHA
5 36 NOS2_MOUSE 7.76 130575 NITRIC OXIDE SYNTHASE, INDUCIBLE
6 36 XDH_MOUSE   7.62 146518 XANTHINE DEHYDROGENASE (EC 1.1
7 37 MDR1_MOUSE 8.51 140994 MULTIDRUG RESISTANCE PROTEIN
8 37 RPB2_ARATH 7.90 135062 DNA-DIRECTED RNA POLYMERASE
9 38 CYA6_MOUSE 8.16 130319 ADENYLATE CYCLASE, TYPE VI (EC
10 38 MSH6_MOUSE 6.61 151076 DNA MISMATCH REPAIR PROTEIN

This message was generated on ExPASy.

B P388/DOX

The SWISS-PROT entries having pI and Mw values in the specified range
for the species MOUSE and the specified keyword:

Rank Score Protein (pI   Mw) Description
==

1 14 MDR1_MOUSE 8.51 140994 MULTIDRUG RESISTANCE PROTEIN
2 19 MDR3_MOUSE 8.94 140755 MULTIDRUG RESISTANCE PROTEIN
3 24 CFTR_MOUSE 8.93 167853 CYSTIC FIBROSIS TRANSMEMBRAN
4 26 MDR2_MOUSE 8.88 140333 MULTIDRUG RESISTANCE PROTEIN
5 28 CO5_MOUSE   6.39 186817 COMPLEMENT C5.
6 29 CYA9_MOUSE 6.85 150954 ADENYLATE CYCLASE, TYPE IX (EC
7 29 PHYD_ARATH 5.81 129302 PHYTOCHROME D.
8 29 CO3_MOUSE   6.39 184179 COMPLEMENT C3.
9 30 A2MG_MOUSE 6.12 134726 ALPHA-2-MACROGLOBULIN 165 KD
10 31 AT7A_MOUSE 6.18 161916 COPPER-TRANSPORTING ATPASE

This message was generated on ExPASy.

99

in the blotting cassette. Take care to remove any trapped air bubbles. Soak the cassette in the tank containing 10 mM CAPS buffer.

7. The membrane was handled with gloves to avoid contamination with skin protein.
8. The blade is washed three times immersing sequentially in three vials of 50% MeCN for each sample, to avoid cross-contamination of samples.
9. It is important for the membrane to be sufficiently wet. The next washing step is inadequate if the membrane is dry. The wet membrane sinks to the bottom of the tube containing 0.01 N HCl solution.
10. As the optimal pH for AQC-derivatization is 8.8, any residual HCl prevents the reaction *(4)*.
11. The excess reagent is destroyed immediately after derivatization
12. Although the resolution for Lys and Ile changes with column aging, it can be improved by changing the column temperature (50°C from 40 min to 60 min).
13. Column maintenance is performed with the same elution program as that for analysis, except that the flow rate is 0.1 mL/min.
14. AQC-Met is easily oxidized in the diluted solution, and is converted to AQC-methionine sulfoxide, which is eluted between Ser and Gly.
15. The concentration of amino acids may differ by more than 10–20-fold. Also, the fluorescence detection loses its linearity, as a result of quenching by high concentration of fluorophore. Therefore, it is better to analyze two levels of sample concentration successively (for example, an intact sample solution [x 1] and a diluted sample [x 10]).
16. For PROPSEARCH in EMBL, the corrected amino acid composition is used for the database search. That is calculated from the experimental amino acid composition divided by the theoretical one using ProtParm tool (http://www.expasy.ch/tools/#primary).

References

1. Shindo, N., Nojima, S., Fujimura, T., Taka, H., Mineki, R., and Murayama, K. (1997) Separation of 18 6-aminoquinolyl-carbamyl-amino acids by ion-pair chromatography. *Anal. Biochem.* **249,** 79–82.
2. Shindo, N., Fujimura, T., Kazuno, N. S., Mineki, R., Furusawa, S., Sasaki, K., and Murayama, K. (1998) Identification of multidrug resistance protein 1 of mouse leukemia P388 cells on a PVDF membrane using 6-aminoquinolyl-carbamyl (AQC)-amino acid analysis and World Wide Web (WWW)-accessible tools. *Anal. Biochem.* **264,** 251–258.
3. Wilkins, M. R., Yan, J. X., and Gooley, A. A. (1998) 2-DE spot amino acid analysis with 9-fluorenylmethyl chloroformate, in *Methods in Molecular Biology*, vol. 112 (Link, A. J., ed.), Humana, Totowa, NJ, pp. 445–460.
4. Cohen, S. A. and Michaud, D. P. (1993) Synthesis of a fluorescent derivatizing reagent, 6-aminoquinolyl-N-hydroxysuccinimidyl carbamate, and its application for the analysis of hydrolysate amino acids via high-performance liquid chromatography. *Anal. Biochem.* **211,** 279–287.

9

Capillary Gas Chromatographic Analysis of Protein and Nonprotein Amino Acids in Biological Samples

Hiroyuki Kataoka, Sayuri Matsumura, Shigeo Yamamoto, and Masami Makita

1. Introduction

Gas chromatography (GC) has been widely utilized for amino acid analysis because of its inherent advantages of high resolving power, high sensitivity, simplicity, and low cost. However, the main problem is the necessity to derivatize the amino acids into less polar and more volatile compounds. Therefore, not only the development of GC systems, but also the concomitant development of suitable derivatives are important factors to accomplish satisfactory GC analysis. An enormous number of approaches for the preparation of volatile derivatives suitable for quantitative analysis of protein and nonprotein amino acids has been reported.

The N-perfluoroacyl alkyl esters including the N-trifluoroacetyl (TFA) methyl esters *(1,2)*, the N-TFA n-propyl esters *(3)*, the N-TFA n-butyl esters *(4,5)*, N-pentafluoropropionyl isopropyl esters *(6)*, the N-heptafluorobutyryl isobutyl esters *(5,7–12)*, and the N-pentafluorobenzoyl isobutyl esters *(13)* have been studied most extensively, some of which have been practically employed for the successful quantitation of amino acids in biological samples. However, it has been recently pointed out that some of these derivatives possess some negative aspects of the derivatization procedures, such as requiring anhydrous conditions, high reaction temperatures, degradation of the amides (glutamine and asparagine) caused by the HCl catalyst and low solubilities of some amino acids in the higher alcohol. The single derivatives including the trimethylsilyl derivatives *(14)* and N,(O)-*tert*-butyldimethylsilyl (*tert*-BDMS) derivatives

From: *Methods in Molecular Biology*, vol. 159: *Amino Acid Analysis Protocols*
Edited by: C. Cooper, N. Packer, and K. Williams © Humana Press Inc., Totowa, NJ

(15–20) have also been reported. Although these methods are nonoxidative, one-step procedure, and will not hydrolyze peptides, these derivatization reactions require anhydrous conditions, high temperatures, and long times. Furthermore, these derivatives give multiple peaks for some amino acids, and some derivatives with higher boiling points tend to decompose on long capillary column. On the other hand, we developed new volatile derivatives, N(O,S)-isobutoxycarbonyl (isoBOC) methyl esters, which were prepared by reaction with isobutyl chloroformate (isoBCF) in aqueous alkaline media, followed by esterification with diazomethane *(21–24)*. The derivatives can be easily, rapidly, and quantitatively prepared without the use of heat and were stable to moisture, and thus appeared to be useful in practice. Subsequently, the modified method based on the preparation of N(O,S)-isoBOC *tert*-BDMS derivatives was reported *(25)*. However, these methods require dual columns for the complete resolution of all amino acids and conversion of arginine to ornithine with arginase.

Recently, an alternative approach for the preparation of the N(O,S)-ethoxycarbonyl ethyl esters with ethyl chloroformate (ECF) was introduced *(26,27)*. Although this method is more simple and rapid by one-step reaction in water-ethanol-pyridine and shows excellent resolution on a capillary column, it is not selective for amino acids because ECF also reacts with other compounds such as fatty acids *(28)*, amines *(29)*, and other organic acids *(30)* under the same reaction conditions, and derivatization of arginine was not achieved. Current methods for the determination of amino acids have also been described in detail (**refs.** *31–36*).

Although our aforementioned method *(21–24)* is relatively time-consuming and requires two packed columns on GC separation, the derivatives could be easily and quantitatively prepared in aqueous media without the use of heat and were stable to moisture. Recently, we have achieved the improvement of our previous method in terms of speed and simplicity by using a capillary column and sonication technique. We have previously reported a simple and rapid method for the determination of protein amino acids base on the preparation of N(O,S)-isoBOC methyl ester derivatives of amino acids by using sonication technique and subsequent GC analysis with hydrogen flame ionization detection (FID) *(37)*. By using the FID-GC method, we have demonstrated that the 22 protein amino acids could be quantitatively and reproducibly resolved as single peaks within 9 min, and the amino acid compositions of proteins *(37)* and protein amino acid contents in serum *(38)* could be rapidly and simply determined. Furthermore, we have demonstrated that the 21 protein amino acids and 33 nonprotein amino acids could be simultaneously resolved within 25 min *(39)* and selectively and sensitively detected by nitrogen-phosphorus selective detection (NPD) *(40)*.

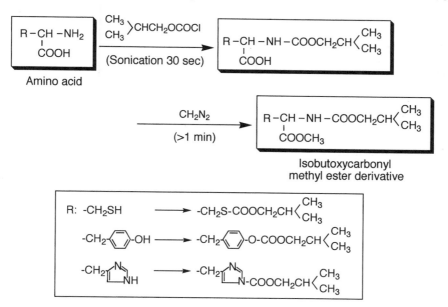

Fig. 1. Derivatization process of amino acids.

By using the NPD-GC method, the protein and nonprotein amino acids in small urine and serum samples *(40,41)* could be directly analyzed without prior clean-up and any interference from other substances. In this chapter, a method for the derivatization and FID- and NPD-GC analyzes of protein and nonprotein amino acids in biological samples is described on the basis of these results, and optimum conditions and typical problems encountered in the development and application of the method are dicussed.

The derivatization process is shown in **Fig. 1**. N(O,S)-isoBOC methyl esters of amino acids can be easily, rapidly, and quantitatively prepared by the reaction with isoBCF by sonication in aqueous alkaline media, followed by esterification with diazomethane. In this derivatization, all amino, imino, imidazole, phenolic hydroxyl, and sulfhydryl groups are substituted with isoBOC groups, whereas alcoholic hydroxyl groups and indole ring nitrogen, as well as amide groups are not. On the other hand, arginine, an exceptional amino acid in our method, is successfully analyzed as the derivative of ornithine after facile treatment with arginase *(37)*. The chief advantage of this method is that the amino acids in aqueous solution can be easily converted into their N(O,S)-isoBOC amino acids without the necessity of heating and without any step to exclude water being necessary prior to derivatization. Furthermore, the derivatives are stable to moisture and can be quantitatively and reproducibly resolved as single peaks using a capillary column.

2. Materials

2.1. Equipment

1. Hewlett-Packard 5890 Series II gas chromatograph equipped with a hydrogen flame ionization detector and a nitrogen-phosphorus detector, an electronic pressure control (EPC) system and a split/splitless capillary inlet system.
2. For the FID-GC analysis, a fused-silica capillary column of crosslinked DB-17 (50% phenyl-50% methylpolysiloxane, J & W, Folsom, CA: 15 m × 0.25-mm id, 0.25-μm film thickness).
3. For the NPD-GC analysis, a fused-silica capillary column of crosslinked DB-17ht (50% phenyl-50% methylpolysiloxane, J & W, Folsom, CA: 20 m × 0.32-mm id, 0.15-μm film thickness).
4. Model RD-41 centrifugal evaporator (Yamato Kagaku, Tokyo, Japan).
5. Pico-Tag workstation (Waters Associates, Milford, MA).
6. Model UT-104 Ultra sonic (39 kHz) cleaner (Sharp, Tokyo, Japan).
7. Pasteur capillary pipet (Iwaki glass No. IK-PAS-5P).

2.2. Reagents

1. Protein amino acids: glycine (Gly), L-alanine (Ala), L-valine (Val), L-leucine (Leu), L-isoleucine (Ile), L-threonine (Thr), L-serine (Ser), L-proline (Pro), L-aspartic acid (Asp), L-glutamic acid (Glu), L-methionine (Met), L-hydroxyproline (Hyp), L-phenylalanine (Phe), L-arginine (Arg), L-lysine (Lys), L-histidine (His), L-tyrosine (Tyr), L-tryptophan (Trp), and L-cystine (Cyt) were purchased from Ajinomoto (Tokyo, Japan); and L-cysteine (Cys), L-asparagine (Asn), and L-glutamine (Gln) from Nacalai Tesque (Kyoto, Japan).
2. Nonprotein amino acids: L-α-aminobutyric acid (α-ABA), α-aminoisobutyric acid (α-AIBA), DL-β-aminobutyric acid (β-ABA), DL-β-aminoisobutyric acid (β-AIBA), β-alanine (β-Ala), DL-norvaline (NVal), DL-norleucine (NLeu), γ-aminobutyric acid (GABA), DL-homoserine (HSer), L-thioproline (TPro), DL-homocysteine (HCys), L-ornithine (Orn) and DL-homocystine (HCyt) were purchased from Nacalai Tesque; ε-amino-*n*-caproic acid (ε-ACA), *p*-aminobenzoic acid (*p*-ABzA), and DL-α-aminopimelic acid (α-APA) from Tokyo Kasei Kogyo (Tokyo, Japan); L-allo-isoleucine (AIle), L-pipecolic acid (PCA), δ-aminolevulinic acid (δ-ALA), DL-α-aminoadipic acid (α-AAA), DL-2,3-diaminopropionic acid (DAPA), DL-2,4-diamino-butyric acid (DABA), L-methionine sulphone (Met-S), DL-δ-hydroxylysine (δ-HLys) and L-cystathionine (CTH) from Sigma (St. Louis, MO).
3. Internal standards: isonipecotic acid and 4-piperidinecarboxylic acid as internal standards (IS) were purchased from Tokyo Kasei Kogyo.
4. Protein samples: bovine serum albumin (Fraction V, 96–99%), hen ovalbumin (Type VII), calf thymus histone (Type II), and calf skin collagen (Type III, acid soluble) were purchased from Sigma.
5. Derivatizing reagents: isoBCF obtained from Tokyo Kasei Kogyo was used without further purification and stored at 4°C. N-Methyl-N-nitroso-*p*-toluenesulpho-

namide and diethyleneglycol mono-methyl ether for the generation of diazomethane *(42)* were obtained from Nacalai Tesquque.

6. Other materials: dithioerythritol (DTE) was obtained from Nacalai Tesque. Peroxide-free diethyl ether was purchased from Dojindo Laboratories (Kumamoto, Japan). Distilled water was used after freshly purification with a Model Milli-Q Jr. water purifier (Millipore, Bedford, MA). All other chemicals were analytical grade.

2.3. Solutions

1. Standard amino acid solutions: three standard stock solutions (each 1 mg/mL), one containing the 22 protein amino acids, the second containing the 22 protein amino acids plus Orn, and the third containing the 21 protein amino acids (except for Arg) plus the 25 nonprotein amino acids; prepared in 0.05 M HCl.
2. Working standard solutions: 1 mg/mL (for FID-GC analysis) or 2 µg/mL (for NPD-GC analysis) of each amino acid were made up freshly as required by mixing of the stock solutions and then dilution with 0.05 M HCl. These working standard solutions are stable at 4°C for at least 2 wk.
3. Internal standard solutions: isonipecotic acid and 4-piperidinecarboxylic acid are dissolved in 0.05 M HCl at a concentration of 1 mg/mL as stock solution. These IS solutions are stable at 4°C for at least 2 wk.
4. Arginase solution: 5 mg of arginase (40 U/mg, Sigma) activated in 4.5 mL of 0.1 M ammonium acetate and 0.5 mL of 0.5 M manganese (II) sulfate at 37°C for 4 h. Centrifuge the solution for 1 min at 2000g, then separate the supernatant and store this solution at –20°C until used. This stored solution of activated arginase can be used without loss of activity for at least 15 d.
5. DTE solution: DTE is used as a 0.5 mM solution in distilled water.

3. Methods
3.1. Derivatization

Derivatization procedure of amino acids is shown in **Fig. 2**.

1. For the FID-GC analysis of protein amino acids, pippette an aliquot of the sample containing 0.2–50 µg of each amino acid and 20 µL of 0.1 mg/mL isonipecotic acid (IS) (if necessary) into a 10 mL Pyrex glass reaction tube with a PTFE-lined screw-cap.
2. For the NPD-GC analysis of protein and nonprotein amino acids, pipet an aliquot of the sample containing 0.02–2 µg of each amino acid and 0.1 mL of 2 µg/mL 4-piperidinecarboxylic acid (IS) (if necessary) into a reaction tube.
3. Add 50 µL of 0.5 mM DTE (*see* **Note 1**) and 0.25 mL of 10% Na_2CO_3 to these solutions and make the total volume up to 1 mL with distilled water.
4. Immediately add 20 µL of isoBCF to the mixture and sonicate for 30 s at room temperature after shaking for a few s by hand (*see* **Notes 2–4**).
5. Extract the reaction mixture with 3 mL of peroxide-free diethyl ether, to remove the excess of reagent, with vigorous shaking for 5–10 s by hand (*see* **Note 5**).

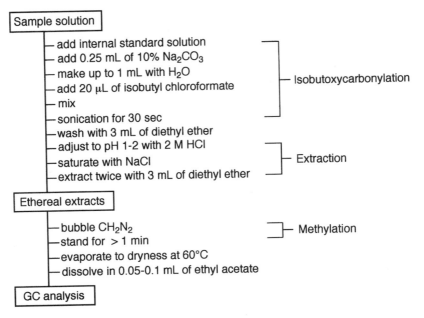

Fig. 2. Shematic flow diagram of the derivatization procedure.

6. Centrifuge at 2000*g* for 30 s, then discard the ethereal extract.
7. Acidify the aqueous layer to pH 1.0–2.0 with 2 *M* HCl and saturate with NaCl. Then extract twice with 3 mL of peroxide-free diethyl ether with vigorous shaking for 5–10 s by hand (*see* **Note 6**).
8. Again centrifuge at 2000*g* for 30 s. Transfer the ether layers into another tube by means of a Pasteur capillary pipet.
9. The pooled ethereal extracts are methylated by bubbling diazomethane, generated according to the microscale procedure *(40)* (*see* **Note 7**). As shown in **Fig. 3**, a stream of nitrogen is saturated with diethyl ether in the first side-arm test tube and then passed through a diazomethane generating solution containing N-methyl-N-nitroso-*p*-toluenesulphonamide, diethylene-glycol monomethyl ether and KOH. The generated diazomethane is carried into the sample tube via the nitrogen stream until a yellow tinge became visible. After standing for 1 min at room temperature, the solvents are removed by evaporation to dryness at 60°C under a stream of dry air. The residue on the walls of the tube is dissolved in 0.05–0.1 mL of ethyl acetate and then 1 µL of this solution is injected into the gas chromatograph by hot-needle-injection technique (*see* **Note 8**).
10. On the other hand, Arg is analyzed as the derivative of Orn after treatment with arginase (*see* **Note 9**). For conversion of Arg into Orn, pipet an aliquot of the solution containing 0.5–10 µg of Arg and 20 µL of 0.1 mg/mL IS (if necessary) into a reaction tube. To this solution, add 0.5 mL of 0.2 *M* Na$_2$CO$_3$-NaHCO$_3$ buffer (pH 9.5) and make the total volume up to 0.7 mL with distilled water. Add

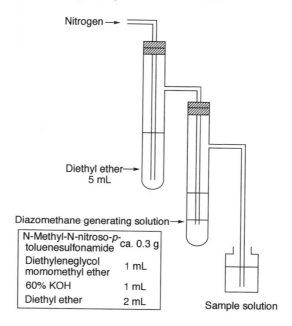

Fig. 3. Apparatus for microscale generation of diazomethane.

the activated arginase solution (50 μL), incubate at 37°C for 10 min, and then derivatize as above (*see* **Note 10**).

3.2. Gas Chromatography

1. FID-GC analysis is carried out with a Hewlett-Packard 5890 Series II gas chromatograph equipped with a hydrogen flame ionization detector and a split injection system. A fused-silica capillary column of crosslinked DB-17 (15 m × 0.25-mm id, 0.25-μm film thickness) is used. The operating conditions are as follows: column temperature program at 30°C/min from 140°C to 290°C and then hold for 5 min; injection and detector temperatures, 300°C; helium carrier gas and make-up gas flow rates, 1 mL/min and 30 mL/min, respectively; split ratio, 50:1.
2. NPD-GC analysis is carried out with a Hewlett-Packard 5890 Series II gas chromatograph equipped with a nitrogen-phosphorus detector (NPD), an electronic pressure control (EPC) system, and a split/splitless capillary inlet system. A fused-silica capillary column of crosslinked DB-17ht (20 m × 0.32-mm id, 0.15-μm film thickness) is used. The operation conditions are as follows: column temperature program at 4°C/min from 120°C to 160°C, then program at 6°C/min from 160°C to 200°C and at 10°C/min from 200°C to 310°C; inlet helium carrier gas flow rate, control at 2 mL/min constant with EPC; split ratio, 10:1. Injection and detector temperatures and make-up gas flow rate are 320°C and 30 mL/min, respectively (*see* **Notes 11** and **12**).

3. A chromatographic blank run (run made with no sample injected) data should be subtracted from the sample run data to remove baseline drift (usually caused by column bleed) using a single-column compensation function. Baseline-corrected data is recorded on the chromatogram.

4. Measure the peak heights of amino acids and the IS, and calculate the peak height ratios of amino acids against the IS to construct calibration curves (*see* **Notes 13–19**).

3.3. Sample Preparation

3.3.1. Hydrolysis of Protein Samples

1. Place an aliquot of the protein sample (10–100 µg) in a 5 × 50 mm glass test tube, and dry in a centrifugal evaporator.

2. Hydrolyze the residue with 0.2 mL of 6 *M* HCl containing 1% phenol (*see* **Note 20**) in the vapor phase for 24 h at 110°C under vacuum in a Pico-Tag workstation *(43)* (*see* **Note 21**).

3. To the resulting hydrolysate, add 20 µL of 0.1 mg/mL isonipecotic acid (IS) and extract the mixture twice with 0.3 mL of distilled water.

4. Transfer the extracts to another reaction tube (10-mL Pyrex glass tube with a PTFE-lined screw-cap). The samples are ready for derivatization and FID-GC analysis (*see* **Note 22**).

3.3.2. Preparation of Urine Samples

1. Collect early morning urine samples from healthy volunteers and process immediately or store at –20°C until used.

2. Urine samples (10–50 µL) are directly used for the derivatization and NPD-GC analysis (*see* **Notes 23–25**).

3.3.3. Preparation of Serum Samples

1. Collect venous blood sample from healthy volunteer in vacutainer tubes and centrifuge at 1600 g for 10 min.

2. Carefully collect the serum layer and process immediately.

3. Serum sample (50 µL) are directly used for the derivatization and NPD-GC analysis without deproteinization (*see* **Notes 23–25**).

4. Notes

1. In order to prevent the oxidation of sulfur amino acids during derivatization, DTE was added to reaction mixture *(44)*.

2. Sonication technique, which can easily mix aqueous solution and oily isoBCF reagent by its vibration action, was useful for acceleration of the reaction.

3. The N(O,S)-isobutoxycarbonylation of amino acids with isoBCF was completed within 15 s by sonication at room temperature. The optimum concentration of Na_2CO_3 was 2.5% as previously described *(21–24)* and the reaction yields for each amino acid were constant with >10 µL of isoBCF.

4. All amino, imino, imidazole, phenolic hydroxyl, and sulfhydryl groups are substituted with isoBOC groups, whereas alcoholic hydroxyl groups, guanidino group, and indole ring nitrogen, as well as amide groups are not.

5. The reaction mixture was washed with diethyl ether under alkaline condition in order to remove the excess reagent. This procedure also serves to exclude amines and phenols both of which often coexist with amino acids in biological samples, as they are derivatized to the corresponding N- or O-isoBOC derivatives, which are soluble in organic solvents under the same conditions as aforementioned *(45–47)*.

6. The resulting N(O,S)-isoBOC amino acids in aqueous layer could be quantitatively and selectively extracted into diethyl ether after acidification to pH 1.0–2.0. In these extractions, peroxide-free diethyl ether should be employed as an extraction solvent to obviate decomposition of sulfur amino acids, and great care should be taken for this solvent when small amounts of these amino acids are to be analyzed by NPD-GC. Furthermore, the ether layers should be collected, taking care to avoid aqueous droplets. It was not necessary to complete draw the ether layer in each extraction.

7. The subsequent methylation of the ethereal extracts could be successfully carried out by bubbling diazomethane. This reaction should be performed in a well-ventilated hood because diazomethane is explosive and toxic.

8. The derivative preparation could be performed within 10 min and the N(O,S)-isoBOC methyl ester derivatives of protein and nonprotein amino acids were very stable under normal laboratory conditions, and no decomposition was observed during GC analysis.

9. Arg requires conversion into Orn with arginase prior to derivatization, because the guanidino group of Arg cannot be derivatized by this method and this amino acid therefore cannot be extracted into diethyl ether because of its polarity.

10. In this reaction, the conversion yield of Arg into Orn was above 93% at each level of Arg from 0.5–10 µg.

11. In a preliminary test for several megabore capillary columns, DB-17 and DB-1701 gave best separation for protein amino acids, although some pairs of amino acids were partially overlapped, e.g., Thr-Ser in DB-17 column, and Leu-Ile and Glu-Met in DB-1701 column *(37)*. Therefore, we tested DB-17 narrow-bore columns at length of 5–20 m to obtain optimum analytical conditions with respect to the separation of amino acids and analysis time. Although the resolution of amino acids increased according to increasing length of column, the analysis time became long. Among the columns tested, 15-m column of DB-17 was proved to be best column for these purposes. As shown in **Fig. 4**, 22 protein amino acids could be completely resolved as single and symmetrical peaks within 9 min on a single capillary column (DB-17, 15 m × 0.25-mm id, 0.25-µm film thickness) *(37)*. However, the separation of protein and nonprotein amino acids were incomplete on this column. Furthermore, high boiling point derivatives were eluted slowly with a broader peak width, and their sensitivities were reduced because of reduction in carrier gas flow rate at higher temperature.

12. In order to solve above problems, we tried to introduce a EPC system and a thin-film-coated and high temperature (max. at 340°C) column DB-17ht (20 m × 0.32-mm id, 0.15-µm film thickness). Of several GC conditions tested for this column, the three-ramp temperature programmes and EPC programs given in **Subhead-**

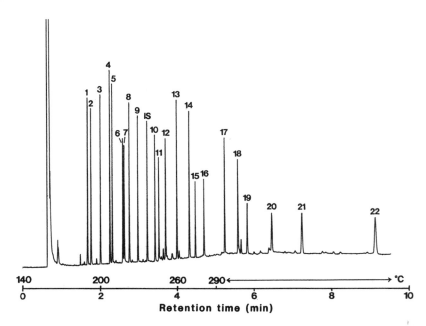

Fig. 4. Standard chromatogram of the N(O,S)-isobutoxycarbonyl methyl ester derivatives of protein amino acids (containing 25 µg of asparagine, glutamine and cystine, and 5 µg of other amino acids) by FID-GC. GC conditions: column, DB-17 (15 m × 0.25 mm ID, 0.25-µm film thickness); column temperature, programmed at 30°C/min from 140°C to 290°C and then held for 5 min; injection and detector temperatures, 300°C; helium carrier gas and make-up gas flow rates, 1 mL/min and 30 mL/min, respectively; split ratio, 50:1. Peaks: 1 = alanine; 2 = glycine; 3 = valine; 4 = leucine; 5 = isoleucine; 6 = threonine; 7 = serine; 8 = proline; 9 = aspartic acid; 10 = glutamic acid; 11 = methionine; 12 = hydroxyproline; 13 = phenylalanine; 14 = asparagine; 15 = cysteine; 16 = glutamine; 17 = ornithine; 18 = lysine; 19 = histidine; 20 = tyrosine; 21 = tryptophan; 22 = cystine; IS = isonipecotic acid (from **ref. *37***, with permission).

ing 3.2. were proved to give the most satisfactory separation of the protein and nonprotein amino acids. As shown in **Fig. 5**, the 21 protein amino acids and the 25 nonprotein amino acids could be well separated as single symmetrical peaks within 28 min on a DB-17ht capillary column, except for δ-HLys, which showed two peaks caused by the allo form present in the standard *(41)*.

13. The reproducibility of sample injection, expressed as percent variation from the mean peak height ratio against the IS, was determined from three independent injections of the standard derivative mixture. For syringe manipulation, a hot-needle-injection technique (the needle inserted into injection zone was allowed to heat up for 3 s prior to the sample injection) was used to prevent the sample discrimination caused by incomplete vaporization and flashback into other parts of the inlet. A split-injection system was proved to be reproducible and the rela-

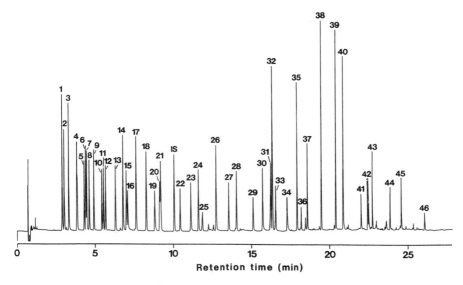

Fig. 5. Standard chromatogram obtained from the N(O,S)-isobutoxycarbonyl methyl ester derivatives of protein and nonprotein amino acids (containing 0.4 μg of asparagine and glutamine, and 0.2 μg of other amino acids) by NPD-GC. GC conditions: column, DB-17ht (20 m × 0.32 mm ID, 0.15-μm film thickness); column temperature, programmed at 4°C/min from 120 to 160°C, then programmed at 6°C/min from 160 to 200°C and at 10°C/min from 200 to 310°C; injection and detector temperatures, 320°C; inlet helium carrier gas flow rate, controlled 2 mL/min constant with EPC; make-up gas flow rate, 30 mL/min; split ratio, 10:1. Peaks: 1 = α-aminoisobutyric acid, 2 = alanine, 3 = glycine, 4 = α-aminobutyric acid, 5 = valine, 6 = β-alanine, 7 = β-aminobutyric acid, 8 = β-aminoisobutyric acid, 9 = norvaline, 10 = leucine, 11 = *allo*-isoleucine, 12 = isoleucine, 13 = norleucine, 14 = γ-aminobutyric acid, 15 = threonine, 16 = serine, 17 = proline, 18 = pipecolic acid, 19 = homoserine, 20 = δ-aminolevulinic acid, 21 = aspartic acid, 22 = thioproline, 23 = ε-aminocaproic acid, 24 = glutamic acid, 25 = methionine, 26 = hydroxyproline, 27 = α-aminoadipic acid, 28 = phenylalanine, 29 = α-aminopimelic acid, 30 = asparagine, 31 = *p*-aminobenzoic acid, 32 = 2,3-diaminopropionic acid, 33 = cysteine, 34 = glutamine, 35 = 2,4-diaminobutyric acid, 36 = homocysteine, 37 = methionine sulphone, 38 = ornithine, 39 = lysine, 40 = histidine, 41 = tyrosine, 42 = d-hydroxylysine, 43 = tryptophan, 44 = cystathionine, 45 = cystine, 46 = homocystine, IS = 4-piperidinecarboxylic acid (from **ref. *41***, with permission).

tive standard deviations ranged from 0.06 to 2.9%. On the other hand, a splitless-injection system could introduced most of sample onto the column and increased sensitivity, but it was proved to reduce reproducibility because of poor refocusing *(39)*. Particularly, the relative responses of high-boiling-point compounds, such as the derivatives of LTH, CTH, Cyt, and HCyt were remarkably reduced. Therefore, we adopted a hot-needle-injection technique in split mode as optimum GC injection technique.

14. The calibration curves for protein amino acids by FID-GC were conducted using isonipecotic acid, which showed a similar behavior to other amino acids during the derivatization and was well separated from other amino acids on a chromatogram as the IS. A linear relationship was obtained with correlation coefficient being above 0.998 in the range 1–50 µg for Asn, Gln, and Cyt, 0.5–10 µg for Arg and 0.2–10 µg for other amino acids. The relative standard deviations for each amino acid in each point were 0.5–7.0% ($n = 3$).

15. The minimum detectable amounts of protein amino acids to give a signal three times as high as the noise under our FID-GC conditions were 0.2–4.0 ng as injection amounts.

16. The contents of Orn and 21 protein amino acids, except for Arg, were calculated from the directly derivatized samples without arginase treatment. On the other hand, Arg content was calculated by subtracting the amount of Orn obtained without arginase treatment from that obtained with arginase treatment *(37)*.

17. NPD-GC is selective and sensitive for nitrogen-containing compounds and the application of this technique to the analysis of amino acids has been reported (8,12,48–52). The NPD-GC system described here was over 10–50 times more sensitive than the FID-GC. Particularly, the increase of sensitivity was remarkable for the nitrogen-rich amino acids such as DAPA, DABA, Orn, Lys, and His *(40)*.

18. The calibration curves for protein and nonprotein amino acids by NPD-GC were conducted using 4-piperidinecarboxylic acid, which showed similar behavior to other amino acids during the derivatization and was well separated from other amino acids on a chromatogram as the IS. Various amounts of protein and nonprotein amino acids ranging from 0.02 to 2 µg were derivatized in the mixture, and aliquots representing 0.2–20 ng were injected into the NPD-GC system. As pointed previously *(12)*, a significant curvature of the detector response was observed by NPD-GC, particularly at a low concentration of amino acid. Therefore, the calibration curves for each amino acid were constructed from both logarithmic plots of the peak height ratios and the amino acid amounts. A linear relationship was obtained with correlation coefficients being above 0.990, in the range 0.04–2 µg for Asn and Gln, and 0.02–1 µg for other amino acids (**Table 1**).

19. The minimum detectable amounts to give a signal-to-noise ratio of 3 under our NPD-GC conditions were 6–150 pg as injection amount (**Table 1**) *(41)*.

20. In order to prevent the oxidation of Tyr and sulfur amino acids during acid hydrolysis, phenol was added to 6 *M* HCl at the concentration of 1%. In this hydrolysis, Trp is generally destroyed, Asn and Gln are quantitatively converted to Asp and Glu.

21. The FID-GC method developed was successfully applied to acid hydrolysate samples of proteins (10–100 µg) *(37)*. Acid hydrolysis of proteins and peptides using a Pico-Tag workstation is not contaminated with reagent because of proceeding in the HCl vapor phase *(43)*.

22. Typical chromatograms obtained from some proteins by FID-GC are shown in **Fig. 6**. The recoveries of total amino acids were 90–104%, and the reproducibility for each amino acid was satisfactory. The amino acid compositions of four well-characterized proteins, bovine serum albumin, hen ovalbumin, calf thymus

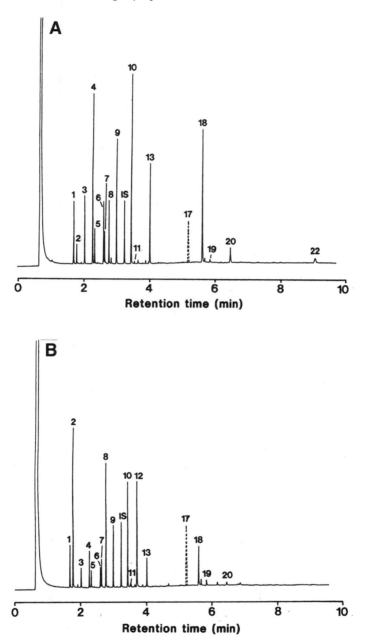

Fig. 6. Typical chromatograms obtained from acid hydrolysates of (**A**) bovine se-rum albumin (100 µg) and (**B**) collagen (100 µg) by FID-GC. GC conditions and peak number: *see* **Fig. 4**. Broken line shows the peak of arginine, which is analyzed as ornithine after arginase treatment (from **ref. *37***, with permission).

Table 1
Linear Regression Data and Detection Limits for Protein
and Nonprotein Amino Acids

Peak No.	Amino acid*	Abbreviation	Regression line[a] Slope a	Intercept b	Correlation coefficient r	Detection limit (pg)
1	α-Aminoisobutyric acid	α-AIBA	0.9981	0.9016	0.9967	9
2	Alanine	Ala	1.0901	0.9456	0.9944	10
3	Glycine	Gly	1.1637	1.0627	0.9952	9
4	α-Aminobutyric acid	α-ABA	1.0942	0.8731	0.9986	10
5	Valine	Val	1.0465	0.7292	09988	11
6	β-Alanine	β-Ala	1.2122	0.8740	0.9987	11
7	β-Aminobutyric acid	β-ABA	1.0833	0.8355	0.9990	11
8	β-Aminoisobutyric acid	β-AIBA	1.1227	0.7655	0.9993	11
9	Norvaline	NVal	1.0645	0.7950	0.9991	11
10	Leucine	Leu	1.0282	0.6375	0.9997	12
11	*allo*-Isoleucine	AIle	1.0163	0.6492	0.9989	12
12	Isoleucine	Ile	1.0185	0.6627	0.9997	12
13	Norleucine	NLeu	1.0670	0.6796	0.9999	12
14	γ-Aminobutyric acid	GABA	1.1886	0.8782	0.9991	11
15	Threonine	Thr	1.2551	0.7440	0.9995	28
16	Serine	Ser	1.4961	0.7727	0.9953	50
17	Proline	Pro	1.0113	0.8217	0.9996	9
18	Pipecolic acid	PCA	1.0078	0.7320	0.9998	11
19	Homoserine	HSer	1.2982	0.6959	0.9992	32
20	δ-Aminolevulinic acid	δ-ALA	1.13 13	0.5799	0.9992	25
21	Aspartic acid	Asp	0.9660	0.6543	0.9997	14
22	Thioproline	TPro	1.2683	0.5937	0.9984	30
23	ε-Aminocaproic acid	ε-ACA	1.0060	0.5151	0.9994	16
24	Glutamic acid	Glu	1.0266	0.6256	0.9998	16

(continued)

histone, and calf skin collagen, determined by this method were good agreement with the literature values. By using this method, protein amino acid contents in serum could be rapidly and simply determined *(38)*.

23. The NPD-GC method developed was successfully applied to the analysis of free amino acids in human urine *(40,41)* and serum *(41)* samples without prior clean-up procedures such as deproteinization, ion–exchange column chromatography, solid phase extraction, and subsequent eluate evaporation.

24. Typical chromatograms obtained from 25 μL of urine and 50 μL of serum by NPD-GC were shown in **Fig. 7**. In urine sample, hippuric acid was observed between TPro and Glu, but it was not overlapped with ε-ACA. Some unknown peaks were observed on the chromatogram, but 37 amino acids detected in the urine could be analyzed without any influence from coexisting substances.

Table 1 (*continued*)
Linear Regression Data and Detection Limits for Protein
and Nonprotein Amino Acids

Peak No.	Amino acid*	Abbreviation	Regression line[a] Slope *a*	Intercept *b*	Correlation coefficient *r*	Detection limit (pg)
25	Methionine	Met	1.3815	0.2873	0.9971	135
26	Hydroxyproline	Hyp	1. 1833	0.8384	0.9987	16
27	α-Aminoadipic acid	α-AAA	1.0597	0.5232	0.9992	24
28	Phenylalanine	Phe	0.9609	0.5718	0.9996	15
29	α-Aminopimelic acid	α-APA	1.0683	0.4584	0.9983	28
30	Asparagiine	Asn	1.3192	0.3684	0.9961	56
31	*p*-Aminobenzoic acid	*p*-ABzA	1.1047	0.6917	0.9980	20
32	2,3-Diaminopropionic acid	DAPA	1.0236	1.0170	0.9984	9
33	Cysteine	Cys	1.1029	0.5802	0.9974	24
34	Glutamine	Gln	1.3119	0.1786	0.9968	150
35	2,4-Diaminobutyric acid	DABA	1.0891	1.0254	0.9991	10
36	Homocysteine	HCys	0.9898	0.1160	0.9931	60
37	Methioninesulphone	Met-S	1.0781	0.7250	0.9933	25
38	Ornithine	Orn	1.0267	1.1324	0.9982	6
39	Lysine	Lsy	1.0214	1.1195	0.9961	6
40	Histidine	His	1.0178	1.0684	0.9963	7
41	Tyrosine	Tyr	1.4371	0.6400	0.9964	64
42	δ-Hydroxylysine	δ-HLys	1.1698	0.6803	0.9934	34
43	Tryptophan	Trp	1.2711	0.8663	0.9951	24
44	Cystathionine	CTH	1.4594	0.7296	0.9936	30
45	Cystine	Cyt	1.2572	0.7120	0.9964	40
46	Homocystine	HCyt	1.2690	0.4084	0.9901	70

[a]$\log y = a \log x + b$: *y*, peak height ratio against the I.S.; *x*, amount of each amino acid (μg); *a*, slope; *b*, intercept. Range: 0.04–2.0 μg for asparagine and glutamine; 0.02–1.00 μg for other amino acids (from **ref. 41**, with permission).

25. To confirm validity of NPD-GC method, known amounts of amino acids were spiked to human urine and serum, and their recoveries were calculated. As shown in **Table 2**, the overall recoveries of these amino acids were 83.5–112.0% and the relative standard deviations were 0.3–14.9% (*n* = 3). The quantitation limits of amino acids in urine and serum samples were approx 0.1–0.4 mg/mL. The intra-assay C.V.s and inter-assay C.V.s for these samples throughout the overall procedure consisting of derivatization and GC analysis were 0.3–8.9% (*n* = 3) and 1.9–15.8% (*n* = 3), respectively.

Table 2
Recoveries of Amino Acids Added to Urine and Serum Samples

Amino acid	Urine				Serum			
	Added (µg/mL)	Amount found (µg/mL)[a] Nonaddition	Amount found (µg/mL)[a] Addition	Recovery (%)	Added (µg/mL)	Amount found (µg/mL) Nonaddition	Amount found (µg/mL) Addition	Recovery (%)
α-AIBA	2	8.61 ± 0.16	10.59 ± 0.80	99.0	1	0.91 ± 0.08	1.89 ± 0.08	98.0
Ala	20	18.48 ± 1.41	38.12 ± 2.17	98.2	10	25.68 ± 0.45	36.07 ± 2.09	103.9
Gly	20	36.49 ± 0.43	57.59 ± 1.62	105.5	10	18.05 ± 0.35	28.23 ± 0.91	101.8
α-ABA	2	0.88 ± 0.07	2.96 ± 0.07	104.0	1	3.15 ± 0.13	4.10 ± 0.04	95.0
Val	2	3.69 ± 0.06	5.59 ± 0.22	95.0	10	29.33 ± 0.74	39.75 ± 1.48	104.2
β-Ala	2	2.45 ± 0.09	4.41 ± 0.20	98.0	1	ND	1.04 ± 0.12	104.0
β-ABA	2	ND[b]	2.13 ± 0.02	106.5	1	ND	1.10 ± 0.09	110.0
β-MBA	20	108.2 ± 2.3	127.8 ± 3.0	98.0	1	ND	0.99 ± 0.03	99.0
NVal	2	ND	2.15 ± 0.05	107.5	1	ND	1.01 ± 0.06	101.0
Lcu	2	3.80 ± 0.07	5.80 ± 0.12	100.0	10	17.52 ± 0.52	26.60 ± 1.68	90.8
Alle	2	ND	2.03 ± 0.10	101.5	1	ND	1.12 ± 0.14	112.0
Ile	2	1.33 ± 0.02	3.41 ± 0.08	104.0	10	10.11 ± 0.20	19.87 ± 1.01	97.6
NLeu	2	ND	2.16 ± 0.12	108.0	1	ND	0.98 ± 0.05	98.0
GABA	2	ND	220 ± 007	110.0	1	ND	1.05 ± 0.05	105.0
Thr	20	11.13 ± 0.19	29.80 ± 0.72	93.4	10	19.85 ± 0.09	29.34 ± 0.74	94.9
Ser	20	19.77 ± 0.99	38.16 ± 1.96	92.0	10	14.03 ± 0.19	23.43 ± 0.76	94.0
Pro	2	0.79 ± 0.06	2.80 ± 0.07	100.5	10	27.65 ± 1.16	37.65 ± 0.71	100.0
PCA	2	ND	1.95 ± 0.06	97.5	1	ND	1.04 ± 0.04	104.0
HSer	2	0.88 ± 0.08	2.91 ± 0.08	101.5	1	ND	0.97 ± 0.11	97.0
δ-ALA	2	1.20 ± 0.07	3.35 ± 0.23	107.5	1	ND	1 07 ± 0.03	107.0
Asp	2	0.80 ± 0.04	2.75 ± 0.16	97.5	1	1.10 ± 0.10	209 ± 008	99.0
TPro	2	1.12 ± 0.07	3.05 ± 0.26	96.5	1	ND	097 ± 0.12	97.0
ε-ACA	2	ND	1.88 ± 0.04	94.0	1	ND	1.07 ± 0.03	107.0

Glu	2	2.28 ± 0.08	4.35 ± 0.14	103.5	10	13.86 ± 0.52	23.51 ± 0.54	96.5
Met	2	1.93 ± 0.22	3.60 ± 0.14	83.5	1	1.00 ± 0.03	1.86 ± 0.04	86.0
Hyp	2	ND	1.83 ± 0.16	91.5	1	1.95 ± 0.07	2.95 ± 0.21	100.0
α-AAA	2	6.31 ± 0.02	8.25 ± 0.30	97.0	1	ND	1.04 ± 0.02	104.0
Phe	2	5.65 ± 0.06	7.75 ± 0.12	105.0	10	10.94 ± 0.39	21.06 ± 0.65	101.2
α-APA	2	3.60 ± 0.12	5.51 ± 0.30	95.5	1	ND	0.99 ± 0.02	99.0
Asn	4	16.40 ± 0.48	20.40 ± 0.22	100.0	2	7.52 ± 0.62	9.66 ± 0.38	107.0
p-ABzA	2	6.16 ± 0.08	8.28 ± 0.16	106.0	1	ND	1.09 ± 0.04	109.0
DAPA	2	ND	2.08 ± 0.18	104.0	1	ND	0.99 ± 0.08	99.0
Cys	2	8.41 ± 0.13	10.39 ± 0.46	99.0	1	0.94 ± 0.04	1.87 ± 0.24	93.0
Gln	40	59.54 ± 1.48	97.02 ± 6.58	93.7	20	100.2 ± 2.8	119.1 ± 3.2	94.5
DABA	2	6.57 ± 0.05	8.43 ± 0.53	93.0	1	0.10 ± 0.01	1.07 ± 0.04	97.0
HCys	20	16.68 ± 0.08	34.87 ± 2.36	91.0	1	ND	0.97 ± 0.08	97.0
Met-S	2	1.68 ± 0.14	3.55 ± 0.26	93.5	1	0.51 ± 0.06	1.48 ± 0.22	97.0
Orn	2	6.60 ± 0.08	8.49 ± 0.13	94.5	10	8.97 ± 0.37	18.55 ± 1.23	95.8
Lsy	20	25.89 ± 0.54	45.17 ± 1.61	96.4	10	34.15 ± 1.37	44.64 ± 2.85	104.9
His	20	158.0 ± 7.0	178.2 ± 9.5	101.0	1	3.55 ± 0.29	4.49 ± 0.52	94.0
Tyr	20	10.33 ± 0.67	31.76 ± 0.76	107.2	1	8.49 ± 0.23	9.48 ± 0.63	99.0
δ-HLys	20	10.71 ± 0.05	28.73 ± 0.94	90.1	1	0.54 ± 0.05	1.46 ± 0.05	92.0
Trp	20	16.39 ± 0.22	36.01 ± 2.39	98.1	1	6.47 ± 0.30	7.38 ± 0.19	91.0
CTH	2	5.01 ± 0.06	7.04 ± 0.31	101.5	1	3.32 ± 0.10	4.23 ± 0.36	91.0
Cyt	20	10.33 ± 0.40	30.05 ± 1.24	98.6	1	0.51 ± 0.04	1.55 ± 0.10	104.0
HCyt	2	6.44 ± 0.04	8.27 ± 0.05	91.5	1	ND	1.08 ± 0.02	108.0

[a]Mean ± SD ($n = 3$).
[b]Not detectable (from **ref. 41**, with permission).

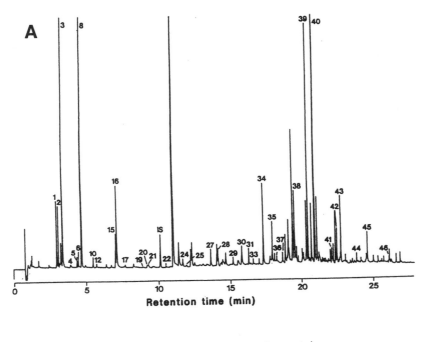

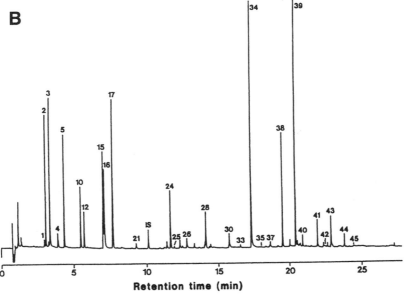

Fig. 7. Typical chromatograms obtained from (**A**) urine (25 µL) and (**B**) serum (50 µL) samples by NPD-GC. GC conditions and peak number: see **Fig. 5**.

References

1. Cruickshank, P. A. and Sheehan, J. C. (1964) Gas chromatographic analysis of amino acids as N-trifluoroacetyl amino acid methyl esters. *Anal. Chem.* **36,** 1191–1197.
2. Darbre, A. and Islam, A. (1968) Gas-liquid chromatography of trifluoroacetylated amino acid methyl esters. *Biochem. J.* **106,** 923–925.
3. Gamerith, G. (1983) Gas-liquid chromatographic determination of N(O, S)-trifluoroacetyl n-propyl esters of protein and non-protein amino acids. *J. Chromatog.* **256,** 267–281.
4. Gehrke, C. W., Kuo, K. C., Kaiser, F. E., and Zumwalt, R. W. (1987) Analysis of amino acids by gas chromatography as the N-trifluoroacetyl n-butyl esters. *J. Assoc. Off. Anal. Chem.* **70,** 160–170.
5. Zumwalt, R. W., Desgres, J., Kuo, K. C., Pautz, J. E., and Gehrke, C. W. (1987) Amino acid analysis by capillary gas chromatography. *J. Assoc. Off. Anal. Chem.* **70,** 253–262.
6. Singh, A. K. and Ashraf, M. (1988) Analysis of amino acids in brain and plasma samples by sensitive gas chromatography-mass spectrometry. *J. Chromatog.* **425,** 245–255.
7. MacKenzie, S. L. and Tenaschuk, D. (1974) Gas-liquid chromatography of N-heptafluorobutyryl isobutyl esters of amino acids. *J. Chromatog.* **97,** 19–24.
8. Chauhan, J., Darbre, A., and Catlyle, R. F. (1982) Determination of urinary amino acids by means of glass capillary gas-liquid chromatography with alkali-flame ionisation detection and flame ionisation detection *J. Chromatog.* **227,** 305–321.
9. Schneider, K., Neupert, M., Spiteller, G., Henning, H. V., Matthaei, D., and Scheler, F. (1985) Gas chromatography of amino acids in urine and haemofiltrate. *J. Chromatog.* **345,** 19–31.
10. Labadarios, D., Shephard, G. S., Botha, E., Jackson, L., Moodie, I. M., and Burger, J. A. (1986) Determination of plasma amino acids by gas chromatography. *J. Chromatog.* **383,** 281–295.
11. MacKenzie, S. L. (1987) Gas chromatographic analysis of amino acids as the N-heptafluorobutyryl isobutyl esters. *J. Assoc. Off. Anal. Chem.* **70,** 151–160.
12. Philpott, M. F. and Van, M. J., der Merwe (1991) Non-linear calibration of a nitrogen-phosphorus detector for the GC determination of amino acids. *Chromatographia* **31,** 500–504.
13. Yeung, J. M., Baker, G. B., and Coutts, R. T. (1986) Simple automated gas chromatographic analysis of amino acids and its application to brain tissue and urine. *J. Chromatog.* **378,** 293–304.
14. Gehrke, C. W. and Leimer, K. (1971) Trimethylsilylation of amino acids. Derivatization and chromatography. *J. Chromatog.* **57,** 219–238.
15. Biermann, C. J., Kinoshita, C. M., Marlett, J. A., and Steele, R. D. (1986) Analysis of amino acids as *tert.*-butyldimethylsilyl derivatives by gas chromatography. *J. Chromatog.* **357,** 330–334.
16. Mawhinney, T. P., Robinett, R. S. R., Atalay, A., and Madson, M. A. (1986) Analysis of amino acids as their *tert.*-butyldimethylsilyl derivatives by gas chromatography and mass spectrometry. *J. Chromatog.* **358,** 231–242.

17. MacKenzie, S. L., Tenaschuk, D., and Fortier, G. (1987) Analysis of amino acids by gas-liquid chromatography as *tert*.-butyldimethylsilyl derivatives. Preparation of derivatives in a single reaction. *J. Chromatog.* **387,** 241- 253.
18. Early, R. J., Thompson, J. R., Sedgwick, G. W., Kelly, J. M., and Christopheason, R. J. (1987) Capillary gas chromatographic analysis of amino acids in blood and protein hydrolysates as *tert*.-butyldimethylsilyl derivatives. *J. Chromatog.* **416,** 15–23.
19. Goh, C. J., Craven, K. G., Lepock, J. R., and Dumbroff, E. B. (1987) Analysis of all protein amino acids as their *tert*.-butyldimethylsilyl derivatives by gas-liquid chromatography. *Anal. Biochem.* **163,** 175–181.
20. Woo, K. L. and Lee, D. S. (1995) Capillary gas chromatographic determination of proteins and biological amino acids as N(O)-*tert*.-butyldimethylsilyl derivatives. *J. Chromatog.* **665,** 15–25.
21. Makita, M., Yamamoto, S., and Kono, M. (1976) Gas-liquid chromato- graphic analysis of protein amino acids as N-isobutoxycarbonylamino acid methyl esters. *J. Chromatog.* **120,** 129–140.
22. Makita, M., Yamamoto, S., Sakai, K., and Shiraishi, M. (1976) Gas-liquid chromatography of the N-isobutoxycarbonyl methyl esters of non-protein amino acids. *J. Chromatog.* **124,** 92–96.
23. Yamamoto, S., Kiyama, S., Watanabe, Y., and Makita, M. (1982) Practical gas-liquid chromatographic method for the determination of amino acids in human serum. *J. Chromatog.* **233,** 39–50.
24. Makita, M., Yamamoto, S., and Kiyama, S. (1982) Improved gas-liquid chromatographic method for the determination of protein amino acids. *J. Chromatog.* **237,** 279–284.
25. Oh, C.-H., Kim, J.-H., Kim, K.-R., Brownson, D. M., and Mabry, T. J. (1994) Simultaneous gas chromatographic analysis of non-protein and protein amino acids as N(O, S)-isobutoxycarbonyl *tert*.-butyldimethylsilyl derivatives. *J. Chromatog.* **669,** 125–137.
26. Husek, P. (1991) Rapid derivatization and gas chromatographic determination of amino acids. *J. Chromatog.* **552,** 289–299.
27. Cao, P. and Moini, M. (1997) Quantitative analysis of fluorinated ethylchloroformate derivatives of protein amino acids and hydrolysis products of small peptides using chemical ionization gas chromatography-mass spectrometry. *J. Chromatog.* A, **759,** 111–117.
28. Husek, P., Rijks, J. A., Leclercq, P. A., and Cramers, C. A. (1990) Fast esterification of fatty acids with alkyl chloroformates. Optimization and application in gas chromatography. *J. High Resolut. Chromatogr.* **13,** 633–638.
29. Husek, P. (1993) Capillary GC analysis of biogenic amines, their precursors and catabolytes after fast derivatization with ethyl chloroformate. *J. Microcol. Sep.* **5,** 101–103.
30. Husek, P. (1995) Simultaneous profile analysis of plasma amino and organic acids by capillary gas chromatography. *J. Chromatog.* B, **669,** 352–357.
31. Rattenbury, J. M. (ed.) (1981) *Amino Acid Analysis*. Ellis Horwood, Chichester, U.K.

32. Zumwalt, R. W., Kuo, K. C. T., and Gehrke, C. W. (eds.) (1987) *Amino Acid Analysis by Gas Chromatography*, vol. I–III. CRC, Boca Raton, FL.
33. Clement, R. E. (ed.) (1990) *Gas Chromatography. Biochemical, Biomedical and Clinical Applications.* Wiley, New York.
34. MacKenzie, S. L. (1981) Recent developments in amino acid analysis by gas-liquid chromatography. *Methods Biochem. Anal.* **27,** 1–88.
35. Labadarios, D., Moodie, I. M., and Shephard, G. S. (1984) Gas chromatographic analysis of amino acids in physiological fluids: a critique. *J. Chromatog.* **310,** 223–231.
36. Walker, V. and Mills, G. A. (1995) Quantitative methods for amino acid analysis in biological fluids. *Ann. Clin. Biochem.* **32,** 28–57.
37. Matsumura, S., Kataoka, H., and Makita, M. (1995) Capillary gas chromatographic analysis of protein amino acids as their N(O, S)-isobutoxycarbonyl methyl ester derivatives. *Biomed. Chromatogr.* **9,** 205–210.
38. Matsumura, S., Kataoka, H., and Makita, M. (1996) Determination of amino acids in human serum by capillary gas chromatography. *J. Chromatog. B* **681,** 375–380.
39. Kataoka, H., Matsumura, S., Koizumi, H., and Makita, M. (1997) Rapid and simultaneous analysis of protein and non-protein amino acids as N(O, S)-isobutoxycarbonyl methyl ester derivatives by capillary gas chromatography. *J. Chromatog. A,* **758,** 167–173.
40. Kataoka, H. (1997) Selective and sensitive determination of protein and non-protein amino acids by capillary gas chromatography with nitrogen-phosphorus selective detection. *Biomed. Chromatogr.* **11,** 154–159.
41. Kataoka, H., Matsumura, S., and Makita, M. (1997) Determination of amino acids in biological fluids by capillary gas chromatography with nitrogen-phosphorus selective detection. *J. Pharm. Biomed. Anal.* **15,** 1271–1279.
42. Schlenk, H. and Gellerman, J. L. (1960) Esterification of fatty acids with diazomethane on a small scale. *Anal. Chem.* **32,** 1412–1414.
43. Bidlingmeyer, B. A., Cohen, S. A., and Tarvin, T. L. (1984) Rapid analysis of amino acids using pre-column derivatization. *J. Chromatog.* **336,** 93–104.
44. Kataoka, H., Tanaka, H., Fujimoto, A., Noguchi, I., and Makita, M. (1994) Determination of sulphur amino acids by gas chromatography with flame photometric detection. *Biomed. Chromatogr.* **8,** 119–124.
45. Makita, M., Yamamoto, S., Katoh, A., and Takashita, Y. (1978) Gas chromatography of some simple phenols as their O-isobutoxycarbonyl derivatives. *J. Chromatog.* **147,** 456–458.
46. Makita, M., Yamamoto, S., Miyake, M., and Masamoto, K. (1978) Practical gas chromatographic method for the determination of urinary polyamines. *J. Chromatog.* **156,** 340–345.
47. Yamamoto, S., Kakuno, K., Okahara, S., Kataoka, H., and Makita, M. (1980) Gas chromatography of phenolic amines, 3-methoxycatecholamines, indoleamines and related amines as their N, O-ethoxycarbonyl derivatives. *J. Chromatog.* **194,** 399–403.
48. Butler, M. and Darbre, A. (1974) Determination of amino acids by gas-liquid chromatography with the nitrogen-sensitive thermionic detector. *J. Chromatog.* **101,** 51–56.

49. Adams, R. F., Vandemark, F. L., and Schmidt, G. J. (1977) Ultramicro GC determination of amino acids using glass open tubular columns and a nitrogen-selective detector. *J. Chromatog.* Sci. **15,** 63–68.

50. Frank, H., Vujtovic-Ockenga, N., and Rettenmeier, A. (1983) Amino acid determination by capillary gas chromatography on chirasil-val. Enantiomer labelling and nitrogen-selective detection. *J. Chromatog.* **279,** 507–514.

51. MacKenzie, S. L. (1986) Amino acid analysis by gas-liquid chromatography using a nitrogen-selective detector. *J. Chromatog.* **358,** 219–230.

52. Buser, W. and Erbersdobler, H. F. (1988) Gas chromatographic determination of amino acids with nitrogen-selective detection. *Zeitsh. Lebens.-Unter. Forsch.* **186,** 509–513.

10

Measurement of Blood Plasma Amino Acids in Ultrafiltrates by High-Performance Liquid Chromatography with Automatic Precolumn O-Phthaldialdehyde Derivatization

Hua Liu

1. Introduction

Amino acid analysis is an important technique that has many applications in biochemical, pharmaceutical, and biomedical fields. Profiling of plasma amino acids is also of great interest in clinical practice. Many diseases are known to be associated with disorders in amino acid metabolism. The analysis of amino acids offers the possibility of genetic prevention in both premarital and prenatal stages for those diseases resulting from inborn errors of metabolism. The plasma amino acid pattern has been also used to follow the course of prolonged dietary treatment *(1,2)*.

For nearly 40 years, amino acid separations have been carried out mainly by an amino acid analyzer by means of ion–exchange chromatography and detected after postcolumn derivatization *(1,3)*. The use of this analyzer has been widely advocated in the past, especially in routine application because of its high reliability *(4)*. However, these analyzes are somewhat laborious, costly, time-consuming, and usually performed on dedicated instruments *(1,5,6)*. In recent years, methods employing precolumn derivatization of amino acids, combined with reversed-phase (RP) high-performance liquid chromatography (HPLC) have gained increasing importance and have partially replaced the classical amino acid analyzer *(6)*. Publications have demonstrated the usefulness of this technique for the determination of amino acids in physiological fluids, and the results achieved by HPLC methods compared favorably with those

From: *Methods in Molecular Biology, vol. 159: Amino Acid Analysis Protocols*
Edited by: C. Cooper, N. Packer, and K. Williams © Humana Press Inc., Totowa, NJ

obtained with the amino acid analyzer (6–14). In comparison with the postcolumn derivatization and ion–exchange chromatographic amino acid analyzers, precolumn derivatization and HPLC methods has the advantages of reducing analysis times, enhancing sensitivity and flexibility, and lowering cost of instrumentation and maintenance (7–16).

Among the methods of precolumn derivatization, O-phthaldialdehyde (OPA) has become the most popular derivatization reagent. In the presence of a thiol compound and at alkaline pH, OPA reacts with primary amino acids and forms highly fluorescent isoindole derivatives. This derivatization procedure is relatively easy and the reaction occurs rapidly at room temperature and in aqueous solution. No laborious purification procedures are required. The OPA derivatives are less polar than the original amino acids and can be well separated from each other by RP-HPLC (17,18). Moreover, the reagent itself does not fluoresce and consequently produces no interfering peaks.

One disadvantage of OPA derivatization has been the lack of stability of the OPA adducts when 2-mercaptoethanol (2-ME) is used as a sulfhydryl reagent (17,19). The reaction products are not stable and have a short half-life, possibly because of a spontaneous intramolecular rearrangement, with sulfur being displaced by oxygen from the ethanolic portion (19). Consequently, time differences between the reaction and injection during a manual procedure may cause significant errors in quantitation. An automated on-line OPA derivatization procedure is described in this chapter. By using an autoinjector, the exact time of each step from the beginning of reaction to injection can be controlled according to an injector program. The precision of the reaction time and sample volume in an automatic procedure eliminates the human errors that may occur during a manual derivatization. In addition, 2-ME was replaced by 3-mercaptopropionic acid (3-MPA) in our experiment with considerable improvement in stability because the stability of isoindoles formed. The other advantages of 3-MPA are that it is nonvolatile and less toxic because of its carboxylic moiety (20).

Another disadvantage of OPA derivatization is that it reacts only with primary amines, so secondary amino acids (imino acids) are not detected (19,21). 9-fluorenylmethyl-chloroformate (FMOC-Cl) is a highly reactive reagent that has been used as an amino-protective group in peptide synthesis. In 1983, Einarsson applied FMOC-Cl to a precolumn derivatization of amino acids (22). It was as sensitive as OPA, but reacted with secondary, as well as primary amine. The derivatives produced were stable and highly fluorescent. This method has been used to measure hydroxyproline, sarcosine and proline in serum, cerebrospinal fluid, and urine (22,23). FMOC-Cl itself is fluorescent and may obscure some amino acid peaks. Therefore, excess FMOC-Cl has to be removed by pentane extraction or reacted with hydrophobic amine 1-aminoadamantane (ADAM) to form an amine-FMOC complex *(24)*. We obviated this problem by using a double-derivative technique (FMOC-Cl is incorporated to this automatic derivatization procedure as a second reagent for the derivatization of secondary amino acids) and using both photodiode array and programmable fluorescence detectors. This procedure has the advantage of different absorption and fluorescent spectra of OPA and FMOC-Cl derivatives. Therefore, both primary and secondary amino acids can be detected simultaneously.

Although there are many techniques available for the analysis of amino acids, and the sample can be analyzed quickly, accurately and sensitively, several precautions need to be taken in order to obtain reliable data on the concentration of plasma amino acids. These include the sample collection, centrifugation, storage conditions, and the deproteinization method. Deproteinization is one of the major problems in the analysis of amino acids in physiological fluid *(25)*. The whole plasma contains soluble peptides and proteins that should be removed from the sample. Otherwise, these substances will clog the chromatographic column, increase instrumental backpressure, and interfere with separation.

The method used to prepare the plasma and to remove the plasma protein has a marked effect on the final results *(1,3)*. The most widely used method of deproteinization is precipitation with 5-sulfosalicylic acid (SSA) followed by centrifugation to remove the precipitated protein *(1)*. We have not had success in using SSA as the deproteinization agent for the analysis of plasma amino acids by an HPLC method with automatic precolumn OPA/3-MPA and FMOC-Cl derivatization *(11)*. When this method was used for the analysis of plasma samples deproteinized by SSA, several major problems were encountered. First, the yield of the derivatization was low in the SSA supernatant. This is probably because the strong acidic nature of SSA inhibits the formation of OPA-amino acid derivatives, which require an alkaline pH. Second, the large peak of deproteinizing agent SSA superimposed the first three amino acid peaks of the chromatogram (*O*-phospho-l-serine (OPS), Asp and Glu). The SSA peaks were higher than 4000, 2000, and 300 mAu at the UV sample wavelengths 230, 260, and 338 nm, respectively. The third problem was the adverse affects of SSA supernatant on the separation and quantitation of other amino acids. Other investigators *(11,18)* have observed similar problems. When ethanol or methanol was used for the deproteinization, the sample was diluted and some of plasma amino acids with lower levels (OPS, Asp, Glu, AABA, and Trp) became undetectable by the UV detector at 338 nm. Other problems were the high level of organic solvent in the injected solution resulting in broad peaks in the early part of the chromatogram, and the increased volatility of the sample made it difficult to store for long periods of time *(11)*.

Ultrafiltration has the advantage of achieving a protein-free sample without adding chemical agents, thus keeping the sample close to a physiological state. It is an attractive alternative to equilibrium dialysis because of the ease and speed with which it can be accomplished. Another potential benefit of ultrafiltration is that the platelets and leukocytes are removed from the plasma, and thus the contamination of amino acids from these blood components can be eliminated. The ultrafiltration methods are not widely applied in the ion–exchange chromatography of amino acids because, for unknown reasons, they decrease the retention time during chromatographic separation and consequently leads to distorted separations of critical pairs of amino acids *(1,3)*.

Ultrafiltration has been chosen for the preparation of protein-free sample in the analysis of amino acid by HPLC methods with precolumn derivatization *(18,26,27)*. The results indicate that ultrafiltration of plasma may replace chemical deproteinization in the HPLC analysis of free amino acids. We investigated several factors that may have an influence on the final results of the ultrafiltration *(11)*. These conditions were then standardized in our procedure. Satisfactory results were achieved for the analysis of plasma amino acids by the automatic precolumn OPA/3-MPA and FMOC-Cl derivatization and RP-HPLC method.

It appears that the combination of: (1) automatic on-line derivatization (2) the improvement in the stability of the reaction by using 3-MPA (3) the use of OPA and FMOC-Cl as dual derivatizing reagents (4) the optimized gradient elution program (5) and the simple mobile phase composition render this method suitable for the quantitative analysis of plasma amino acids in a clinical laboratory. This procedure can yield more rapid and sensitive results than that of classic amino acid analyzer, with comparable accuracy and precision. We have used this method for the determination of plasma amino acids in more than 3000 plasma samples during a period of 10 yr with satisfactory results (*see* **Notes 1** and **2**).

2. Materials

2.1. Equipment

1. A Hewlett-Packard HP 1090 *M* series HPLC system (*see* **Note 3**). The system consists of a DR 5 solvent delivery system with 3 solvent channels, a variable volume autoinjector and an autosampler, an HP 1040A photo-diode array UV detector, and HP 1046 programmable fluorescence detector (*see* **Note 4**).
2. An HP 79994A analytical workstation for data processing.
3. Two HP Hypersil-ODS 5-μm columns (100 × 2.1 mm).
4. One guard column (20 × 2.1 mm) (*see* **Note 5**).

2.2. Reagents

Water, methanol, and acetonitrile were HPLC grade (Curtin Matheson Scientific, Inc., Houston, TX). Other chemicals used were analytical grade including sodium acetate, glacial acetic acid, boric acid, sodium hydroxide, *O*-phthaldialdehyde, 3-mercaptopropionic acid, and 9-fluorenylmethyl-chloroformate (all from Sigma Chemical Company, St. Louis, MO).

2.3. Solutions

1. Standard solution: An amino acid standard solution containing 29 amino acids was prepared by adding crystalline *O*-phospho-L-serine (OPS), reduced glutathione (GSH), asparagine, glutamine, citrulline, taurine, α-amino-N-butyric acid (AABA), tryptophan, ornithine, hydroxy-proline (Hyp), and sarcosine to a 50% methanol solution. Then this solution (containing 11 amino acids) was mixed with a commercial AA-S-18 Amino Acid Standard Solution (from Sigma). Nor-Valine (N-Val) can be used as an internal standard (United States Biochem Corp, Cleveland, OH). The concentration of working standard was 250 µmol/L for each amino acid. The mixture solution was kept in –76°C (*see* **Note 6**).
2. OPA derivatization reagent: The OPA derivatization reagent was prepared by dissolving 3 mg of OPA in 50 µL of methanol, adding 450 mL of sodium borate buffer (0.5 mol/L, pH 10.2) and 5 µL of 3-MPA. Borate buffer was prepared from 0.5 *M* boric acid solution adjusted to pH 10.2 with 5 *M* sodium hydroxide solution. This OPA solution was placed in an amber crimp top vial with a silicone rubber PTFE-coated cap and kept in the dark at –20°C. Fresh solution was prepared each week.
3. FMOC derivatization reagent: FMOC-Cl solution was prepared by dissolving 1.29 mg of FMOC-Cl in 1 mL of acetonitrile and stored at –20°C (*see* **Note 7**).
4. Mobile phase solution: The sodium acetate buffer (0.1 *M*) in mobile phase was prepared by dissolving sodium acetate in HPLC-grade water and titrating to pH 6.8 with glacial acetic acid. The buffer was then diluted to 0.015 *M* and 0.01 *M* for mobile phase solutions A and C, respectively (*see* **Note 8**). The mobile phases were filtered by passing through a 0.45-µm Durapore membrane filter (Millipore Inc., Milford, MA) and continuously degassed by helium.
5. Blood sample: Blood samples were collected from subjects into a heparinized vacutainer tube (*see* **Note 9**). The plasma samples were ultrafiltrated by using the Centrifree System (Amicon, Beverly, MA).

3. Methods

1. Collect 1.5 mL of venous blood samples from subjects by venipuncture between 7 AM to 9 AM (*see* **Note 10**) after overnight fasting and put into a vacutainer tube containing heparin (*see* **Note 11**). Centrifuge the blood samples at 2000*g* for 15 min at 10°C (*see* **Note 12**) and remove the plasma.
2. Ultrafiltrate the heparinized plasma samples by using the commercially available Centrifree System (*see* **Notes 13** and **14**). During the ultrafiltration, the sample is

Table 1
Injector Program[a]

Line#	Function	Amount		Vial No.	Reagent
1	Draw :	0.0 µL from	:	Vial#:4	(Water)
2	Draw :	2.5 µL from	:	Vial#:5	(OPA)
3	Draw :	0.0 µL from	:	Vial#:4	
4	Draw :	2.5 µL from	:	Vial#:X	(Sample)
5	Mix :	5.0 µL cycles 2			
6	Draw :	0.0 µL from	:	Vial#:4	
7	Draw :	1.0 µL from	:	Vial#:8	(FMOC-Cl)
8	Mix :	6.0 µL cycles 2			
9	Wait :	2.5 min			
10	Inject				

[a]Reprinted from **ref. 8**, p. 3326 by courtesy of Marcel Dekker, Inc.

Table 2
Time Table for Gradient Elution

Time (Minute)	Solvent A%	B%	C%
0.05	100	0	0
15.00	60	40	0
18.50	57.5	42.5	0
22.00	45	55	0
25.00	0	0	100
30.00	0	0	100

Solvent A: 0.015 M NaAc buffer (pH 6.8).
Solvent B: Methanol.
Solvent C: 0.010 M NaAc buffer (pH 6.8).
Reprinted from **ref. 8**, p. 3327 by courtesy of Marcel Dekker, Inc.

deproteinized by filtration of plasma through the ultrafiltration membrane. The protein is retained by the membrane, whereas the ultrafiltrate (containing free amino acids) pass through and collect in the filtrater cup. Put a volume of 0.4 mL of plasma in the sample reservoir, and then place the device in a centrifuge with a 45° fixed-angle rotor (*see* **Note 15**). About 80 µL of ultrafiltrates should be collected after centrifugation at 1000g for 15 min. Store the ultrafiltrates at –80°C until analyzed (*see* **Note 16**).

3. To perform the automatic precolumn derivatization procedure using an injector program (**Table 1**), place the OPA-reagent at vial number 5 and the FMOC-Cl reagent at vial number 8. First, 2.5 µL of the OPA reagent is drawn into the sample loop. Then, 2.5 µL of sample is drawn into the loop and mixed with the OPA reagent. Finally, 1 µL of FMOC-CL reagent is drawn and mixed in the sample

loop. After waiting 2.5 min for the reaction, a total 6 μL of sample and reagents mixture is injected into the column for gradient elution (*see* **Note 17**). After drawing from sample or reagents, the needle of the injector should always be dipped into 1 mL of water (vial number 4) for cleaning.

4. The mobile phases are 0.015 M sodium acetic buffer (pH 6.8, for solvent A), methanol (solvent B) and 0.015 M of sodium acetic buffer (pH 6.8, for solvent C) (*see* **Notes 18** and **19**). The separation of amino acids in both the standard solution containing 29 amino acids or plasma samples is carried out by a gradient elution according to a chromatographic time-table (**Table 2**). The flow-rate is 0.3 mL/min and the stop time is 30 min after the injection. Place two HP Hypersil-ODS 5-μm columns (100 × 2.1-mm id) in series in a thermostatically controlled column compartment preceded by a guard column (20 × 2.1-mm id). The column temperature should be maintained at 40°C for the separation of amino acids (*see* **Note 20**).

5. For the detection of amino acid derivatives, set the photo-diode array detector at three sample wavelengths: 338, 266, and 230 nm with bandwidths of 10, 4, and 4 nm, respectively (*see* **Note 21**). The reference wavelength is 550 nm with a bandwidth of 100 nm. The initial parameters for the fluorescence detector are excitation wavelength (Ex) 230 nm and emission wavelength (Em) 450 nm. Twenty minutes after the injection, change the Ex and Em to 260 nm and 315 nm, respectively, for the determination of secondary amino acids proline and hydroxproline (*see* **Note 22**).

6. Amino acid peaks are identified with reference to retention times of standard amino acids injected. Coinjection of standard amino acids and the plasma samples may be needed to identify the amino acid peaks in some samples. Chromatograms of an amino acids standard mixture and a representative plasma sample (*see* **Note 11**) are demonstrated in **Figs. 1** and **2**. These chromatograms showed a satisfactory separation of 29 primary and secondary amino acids.

7. The precision of analysis is observed from the reproducibility of the peak areas of eight consecutive injections of 29 amino acid mixture solution. The results are listed in **Table 3**. The coefficients of variation for peak areas ranged from 0.78% to 2.92%, with a mean of 1.73% ± 0.67% SD (*see* **Note 23**). The high precision of this method would allow analysis without an internal standard for quantitation. The amino acid standard should be first analyzed twice for calibration and thereafter every eighth analysis in an automated series.

4. Notes

1. The levels of 25 plasma amino acids from 75 boys and 85 girls are listed in **Table 4**. In general, amino acid levels of boys were higher than that of girls. The mean values of aspartic acid, methionine, isoleucine, and hydroxy-proline in boys were significantly higher than that in girls ($p < 0.01$). The girls had significant higher level of histidine than boys ($p < 0.001$). Compared with boys below 6 yr old ($n = 23$), boys over 6 yr old ($n = 52$) showed significant higher value of glycine, threo-

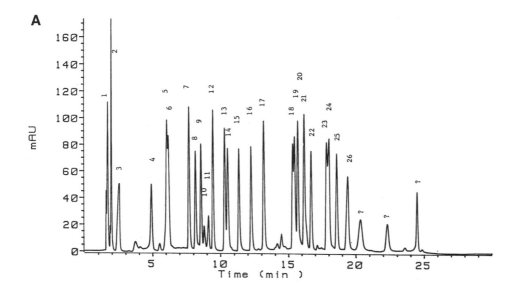

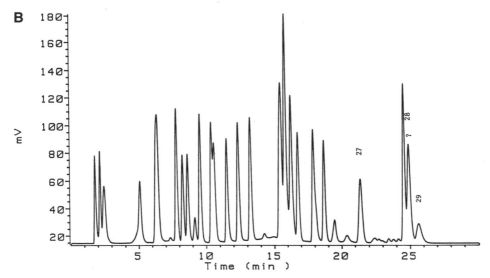

Fig. 1. Typical chromatograms showing the separation of 29 OPA/3-MPA and FMOC-Cl derivatized standard amino acids (500 µmol/L). (**A**) Signal from UV detection at 338 nm. (**B**) Signal from Fluorescence detection. For chromatographic conditions, *see* **Subheadings 2** and **3**. Peaks: 1 = *O*-phospho-L-serine, 2 = Aspartic Acid, 3 = Glutamic Acid, 4 = Glutathione (reduced), 5 = Asparagine, 6 = Serine, 7 = Glutamine, 8 = Glycine, 9 = Threonine, 10 = Histidine, 11 = Cystine, 12 = Citrulline, 13 = Taurine, 14 = Alanine, 15 = Arginine, 16 = Tyrosine, 17 = Alpha-amino-*N*-butyric Acid, 18 = Methionine, 19 = Valine, 20 = Nor-Valine, 21 = Tryptophan, 22 = Phenylalanine, 23 = Isoleucine, 24 = Ornithine, 25 = Leucine, 26 = Lysine, 27 = Hydroxy-proline, 28 = Sarcosine, 29 = Proline.

Reprinted from **ref.** *8*, p. 3328 by courtesy of Marcel Dekker, Inc.

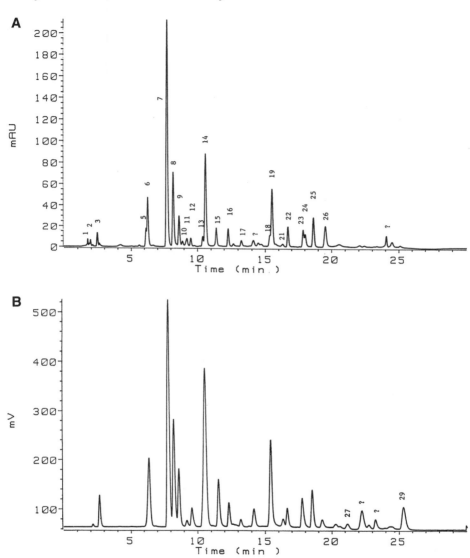

Fig. 2. Chromatograms of a representative plasma sample under the identical experimental conditions as in **Fig. 1**. (**A**) Signal from UV detection at 338 nm. (**B**) Signal from fluorescence detection. Peak numbers are identified in **Fig. 1A** and **B**.

Reprinted from **ref. 8**, p. 3329 by courtesy of Marcel Dekker, Inc.

nine, histidine, taurine, alanine, arginine, tyrosine, valine, phenylalanine, isoleucine, ornithine and leucine. For girls over 6 yr old ($n = 56$), only threonine, histidine, and alanine were significantly higher than that for girls below 6 yr old ($n = 29$) (*8*).

Table 3
Reproducibility of Peak Areas for Standard Amino Acids

Amino Acid[a]	Mean (n = 8)	SD	CV%
OPS	527.6	5.47	1.03
Asp	528.1	4.13	0.78
Glu	545.4	15.49	2.84
GSH	532.4	13.69	2.57
Asn	581.9	16.58	2.84
Ser	694.3	8.96	1.29
Gln	676.1	8.99	1.32
Gly	498.8	9.97	1.99
Thr	536.4	10.78	2.00
His	106.0	1.59	1.50
Cys	170.8	3.94	2.30
Cit	637.7	4.88	0.76
Tau	582.4	5.76	0.98
Ala	589.0	5.22	0.88
Arg	554.4	7.37	1.32
Tyr	553.7	8.14	1.47
AABA	707.5	13.78	1.94
Met	548.4	10.52	1.91
Val	660.3	7.69	1.16
N-Val	829.4	12.16	1.46
Trp	638.4	16.49	2.58
Phe	535.1	11.55	2.15
Ile	646.1	18.51	2.86
Orn	540.6	15.80	2.92
Leu	545.1	10.01	1.83
Lys	587.8	9.54	1.62
Hyp	1469	28.91	1.91
Sar	2427	20.85	0.85
Pro	645.9	8.86	1.37

[a]OPS = O-phospho-L-serine, Asp = Aspartic Acid, Glu = Glutamic Acid, GSH = Glutathione (reduced), Asn = Asparagine, Ser = Serine, Gln = Glutamine, Gly = Glycine, Thr = Threonine, His = Histidine, Cys = Cystine, Cit = Citrulline, Tau = Taurine, Ala = Alanine, Arg = Arginine, Tyr = Tyrosine, AABA = Alpha-amino-N-butyric Acid, Met = Methionine, Val = Valine, N-Val = Nor-Valine, Trp = Tryptophan, Phe = Phenylalanine, Ile = Isoleucine, Orn = Ornithine, Leu = Leucine, Lys = Lysine, Hyp = Hydroxy-proline, Sar = Sarcosine, Pro = Proline.

2. The plasma amino acid levels of 160 children obtained by this method showed a good agreement with the reference values reported previously by using ion–exchange chromatography *(28)*. The findings about the difference between amino acid levels of boys and girls and between the children with different age were

Table 4
Free Plasma Amino Acid Levels (mmol/L) of 160 Children

Amino Acid[a]	Girls ($n = 85$)			Boys ($n = 75$)		
	Mean	±	SD	Mean	±	SD
Asp	6.19		1.86	7.62[b]		2.13
Glu	39.28		13.95	48.23		17.84
Asn	53.99		10.16	56.53		10.92
Ser	126.30		23.57	133.50		28.79
Gln	532.80		82.00	540.70		93.75
Gly	242.50		47.29	262.60		59.35
Thr	142.40		27.86	146.20		28.87
His	96.09[c]		16.51	85.07		16.95
Cys	88.96		14.91	87.21		15.29
Cit	24.29		4.10	26.00		4.15
Tau	41.57		7.73	44.35		11.53
Ala	423.30		82.15	425.10		96.85
Arg	95.05		18.18	100.60		19.52
Tyr	74.88		15.45	73.99		13.51
AABA	18.23		3.52	18.66		3.93
Met	23.88		2.60	25.17[b]		3.65
Val	221.80		39.27	235.20		49.11
Trp	5.67		2.73	6.38		3.33
Phe	64.94		11.15	65.32		5.96
Ile	66.36		15.35	74.28[b]		19.63
Orn	42.11		12.76	45.03		14.53
Leu	137.90		25.92	144.30		32.03
Lys	130.00		25.99	126.10		26.91
Hyp	18.96		4.20	21.34[b]		5.65
Pro	153.00		42.24	149.90		45.66

[a]Amino acid abbreviations as in **Table 3**.
[b]Significantly higher than girls ($p < 0.01$).
[c]Significantly higher than boys ($p < 0.001$).
Reprinted from **ref. 8**, p. 3332 by courtesy of Marcel Dekker, Inc.

also similar to the results observed by Armstrong and Stave *(28,29)*. Patients suffered from hyperlysinemia, nonketotic hyperglycinemia, disorders of branched-chain amino acid metabolism, and argininosuccinase deficiency have been detected by this procedure.

3. This instrument is a low-pressure gradient system. A high-pressure gradient system can be used for the analysis. We recommend that the buffers in a mobile phase be premixed with organic solvents (5% to 10% buffer in organic solvent and 5% to 10% organic solvent in buffer solution), and thorough degassed be performed when a high-pressure gradient system is being used.

4. The contribution to dispersion by the instrument itself must be minimized. Two detectors should be physically close each other. The internal diameter of the capillaries should be as small as possible.

5. When contamination was detected (baseline drift during the gradient elution) or separation deteriorated, and the change of guard column could not correct the problem, the first column (next to the guard column) was discarded, the second one moved to the first position, and a new column was added at second position. Under a normal situation, the guard or analytical column can be used for analysis up to 100 or 500 plasma samples, respectively.

6. Other individual amino acid (such as argininosuccinic acid) can be added to this amino acid standard mixture for peak identification and quantitation.

7. The purity of the reagent was important for extending detection limit. A previously unused and aged reagent (> 24-h old) was necessary for a higher sensitivity level detection *(21)*. The sensitivity depended mainly on the ability to eliminate or subtract background levels and reduce the interfering substances present in solvents and reagents.

8. It is recommended that the concentration of the buffer solutions in the mobile phase be less than 0.1 M to avoid the potential precipitation during the gradient elution.

9. A siliconized tube is recommended to prevent blood platelets becoming activated with the resultant release of Tau and phosphoethanolamine (PEA). Hemolysis should also be prevented because it may lead to false increases in the concentrations of Asp, Glu, Tau, and PEA *(4)*.

10. The time of blood withdrawal and the relation to dietary intake may be another reason that may cause varying results between laboratories. If it is possible, the blood sample should be collected between 8 and 9 AM before breakfast.

11. Many analytical factors could result in variation in the plasma amino acid profile. These include: (1) sample collection (2) centrifugation and ultrafiltration (3) storage conditions (4) contamination by platelets and leukocytes or hemolysis (5) shift in baseline during chromatographic analysis, and (6) interference or overlap of one or more amino acids. The physiological factors such as circadian rhythm, protein intake, pregnancy, nutritional status of the subject, physical activity before blood collecting, menstrual cycle, sex, and age differences and medication that interferes with the analysis may also have influence on plasma amino acids. All of these factors should be controlled to assess with certainty the clinical data under various pathological conditions.

12. Although most of the amino acids remained stable for up to 30 min at room temperature, delayed deprotenization may cause an increase of Asp, Glu, Gly, Ala, and Orn, whereas there is a decrease of Asn, Gln, His, and 3-MH *(2)*. When deproteinization of physiological samples cannot be done immediately after centrifugation, it is recommended to store the specimens at temperature lower than –18°C to prevent further hydrolysis of protein. In particular, the amino acids Asp and Glu increased markedly when the samples were not frozen immediately and not stored at –68°C or lower.

13. In dealing with physiological fluid samples, the protein should be completely removed from the sample prior to analysis. Direct injection onto RP columns are known to affect the retention behavior of the solutes and cause increase back pressure by irreversible adsorption of protein to the stationary phase.

14. Precaution should be taken when this method is used for some samples with higher protein concentration, such as blood cell lysates or tissue extracts. It has been reported that for the lipemia samples, the time required to filter sufficient sample was very variable and occasionally no filtrate could be obtained, presumably because the membrane pores got blocked. The alternative method is chemical precipitation by using various acids, bases, or organic solvents. Precaution should be taken to avoid hydrolysis, degradation, and oxidation of amino acids in selecting any such procedure for the deproteinization. The main factors to be considered are the amount of protein to be removed, the comparability or interference with the chromatographic analysis, the thermal and chemical stability, the solubility of the analytes, and finally, the cost, time, and labor. The use of 24–30% SSA to deprotein 500-µL plasma samples (w/v) has been reported yield consistent results with high recoveries of most amino acids *(2,4)*. Ethanol or acetonitrile (ACN) can also be used as precipitants (ACN or ethanol to plasma ratio 2–4:1, v/v) and has the advantage of resulting in full recovery of total Trp, whereas using SSA precipitation cannot completely recover this amino acid *(4)*.

15. Several factors such as rotor, membrane, and the time of the ultrafiltration may have an influence on the final results of the ultrafiltration *(11)*. These factors were investigated in our experiment. The recovery rates of amino acid in ultrafiltrates obtained by fixed-angle rotor were higher than that by swinging-bucket rotor. This may be caused by the different polarization control between fixed-angle and swinging-bucket rotors. The use of a fixed-angle rotor provides polarization control. The angle counteracts the buildup of retained protein at the membrane surface, because this dense layer slides outward and accumulates at the edge of membrane. In a swinging-bucket rotor, the polarization layer is compacted over the entire membrane surface, restricting the passage of solute and solvents through the membrane. Our experiment suggests that the polarization occurring during the ultrafiltration may be one of the important factors that can influence the recovery rate of amino acids. This might be the reason why some other authors could not achieve satisfactory results by ultrafiltration, because it was difficult to control the polarization when ultrafiltration was performed under nitrogen pressure, or a syringe was used as the driving force for the ultrafiltration.

16. The recovery from ultrafiltration and its reproducibility of standard amino acids were tested by 10 aliquots of amino acid standards: 5 aliquots were analyzed by HPLC before ultrafiltration, another 5 aliquots were analyzed after ultrafiltration. The results showed that the recovery of all of the standard amino acids was excellent, ranging from 95–102% *(11)*. The accuracy of measurement was tested by adding a known quantity of amino acid standards to a plasma sample, then the sample was ultrafiltrated and derivatized for analysis. The analytic recovery rate

Table 5
Recovery of Amino Acid Standard Added to Plasma

Amino Acid[a]	Plasma alone (μmol/L)	Plasma+250 mmol/L (expected)	Actual	Recovery Rate %
Ops	32.65	282.65	271.87	97
Asp	18.98	268.98	273.34	101
Glu	28.76	278.76	290.15	104
GSH	—	250.00	80.73	32
Asn	35.89	285.89	284.52	100
Ser	128.27	378.27	401.06	106
Gln	538.74	788.74	757.19	96
Gly	219.74	469.74	465.04	100
Thr	132.14	382.14	384.32	100
His	31.71	281.71	290.16	103
Cys	64.12	314.12	307.83	98
Cit	18.36	268.36	287.14	107
Tau	38.85	288.85	291.73	101
Ala	376.32	626.32	613.79	98
Arg	103.50	353.50	352.22	100
Tyr	62.38	312.38	303.01	97
AABA	29.09	279.09	287.46	103
Met	19.60	269.60	258.82	96
Val	170.04	420.04	407.73	97
N-Val	—	250.00	252.36	101
Trp	14.26	264.26	214.05	81
Phe	49.82	299.82	297.42	99
Ile	60.21	310.21	313.08	100
Orn	50.18	300.18	285.17	95
Leu	100.92	350.92	349.15	100
Lys	137.24	387.24	385.69	99
Hyp	24.49	274.49	277.23	101
Pro	180.86	430.85	439.46	102

[a]Amino acid abbreviations as in **Table 3**.
Reprinted from **ref. 8**, p. 3331 by courtesy of Marcel Dekker, Inc.

of each amino acid was calculated after HPLC quantitation (**Table 5**). It should be mentioned that when standard amino acids were added to the plasma, Trp and GSH (a peptide) showed poor recovery (81% and 32%, respectively), whereas all of the other amino acids remained at a similar recovery. The reason was unknown. Changing the membrane in the ultrafiltration system to a membrane with higher molecular weight cutoff might be helpful for the improvement of the recovery rate of Trp and GSH.

The multiplier in calibration form was set at 1.19 for the quantitation of plasma tryptophan. No effort has been made for the determination of GSH in plasma.

17. Although the stability of OPA/3-MPA derivative is better than OPA/2-ME derivative, the OPA/3-MPA derivative of various amino acids does not have the same stability *(8)*. When the waiting time was increased beyond 2.5 min, the stability of various amino acids was different. The UV absorbency of GSH and OPS increased, whereas asparagine, glutamine, glycine, histidine, taurine, ornithine, and lysine decreased. Others were unchanged. Therefore, the precise control of reaction time is still important in the derivatization using OPA/3-MPA as reagents. The results from our experiment indicated that OPA/3-MPA derivatives were relative stable during the waiting time from 0.5 to 5 min *(8)*.

18. When analyzing physiological samples with interested amino acids covering a broad retention range, it is necessary to modify the elution conditions during the analysis to optimize the separation. The combinations of the organic solvents and buffer solution; the pH and ion concentration of the buffer in the mobile phase; the patterns of gradient elution, and the different column and column temperature should be investigated to optimize the separation.

19. Comparison of buffer pH from 6.4 to 7.2 was made in our experiment with all other conditions held constant *(8)*. The pH at 6.8 appeared to give optimal separation. The pH of the buffer in the mobile phase should be close to neutral. A lower pH of the buffer would result in weaker UV absorbency and fluorescent intensity.

20. Maintaining a stable column temperature is critical to the reproducibility of retention time, in addition to improving the separation of amino acids. When the column temperature was kept at 40°C, the coefficients of variation for retention time ranged from 0.02% to 0.97%, with a mean of 0.26%+0.21 SD. Column temperature was also tested for its effect on separation. When column temperature was maintained at 30°C, the separation of amino acids at first half of the chromatogram was better than that at temperature 35°C or 40°C, but the separation of second half was not satisfactory. Increasing the temperature to 40°C gave better separation for amino acids eluted later and the overall results were the best, together with a lower-column pressure. We therefore set the column temperature at 40°C for the analysis *(8)*.

21. The OPA/3-MPA and FMOC-Cl derivatives were detected in our experiment by a photo-diode array detector at three different sample wavelengths. The intensity of the signals, the baseline noises, and interference were different at various wavelength settings. At sample wavelengths of 338 nm and 266 nm, we compared two reference wavelengths at 390 nm and 550 nm. The use of 550 nm as reference wavelength produced more stable baseline than that at 390 nm. Therefore, 550 nm was used as reference wavelength for the photo-diode array detector. Under the chromatographic conditions we had chosen, there was no interfering peak on either UV or fluorescence signals during a blank solvent gradient elution. When HPLC water was used as a blank sample, there was no peak interfering with amino acid peak at UV sample wavelength 338 nm, and fluorescence signals. But at UV 266 nm, and especially at 230 nm, several peaks were

observed that were large enough and close enough to interfere with about 10 amino acid peaks.

Although detection at UV 230 nm was more sensitive than at 338 nm, the interference could seriously compromise the quantitation. For this reason, UV 338 nm is recommended as the main sample wavelength for the quantitation of primary amino acids whereas wavelength 266 nm is recommended for the FMOC-Cl derivatives of secondary amino acids. Extremely high sensitivity is not required for the analysis of plasma amino acids, because the limitation is not sample size, rather the amount of plasma required for deproteinization procedure (such as ultrafiltration).

22. For the fluorescence detector we used to detect OPA and FMOC-Cl derivatives, both 230 nm and 340 nm were evaluated as Ex for OPA primary amino acid derivatives. The results showed that the detection of OPA primary amino acid derivatives at Ex 230 nm gave a response over seven times stronger than that at 340 nm, with an acceptable baseline. For the detection of FMOC-Cl secondary amino acid derivatives, three excitation wavelengths of 254, 260, and 266 nm were compared. The results showed 260 nm had the strongest signal. Therefore, we set Ex 230 nm and Em 450 nm during the first 20 min for the measurement of OPA derivatives of primary amino acids, then switched to Ex 266 nm and Em 315 nm for the determination of FMOC-Cl derivatives of secondary amino acids. The hydroxyproline, sarcosine, and proline were all eluted after 20 min in our chromatographic system.

23. Analysis of physiological samples with different amino acid concentrations is required during the practical applications. The linear relationship of amino acid concentrations in the ultrafiltrates was investigated. The linearity of response was estimated by injecting derivatized amino acids with different concentrations and constructing regression equations for UV and fluorescence response-concentration curves. The linear relationship between the concentration and peak areas of each standard amino acid was determined by analyzing the standard amino acid mixture at concentrations ranging from 31.25 to 500 μmol/L ($n = 5$, by serial dilution). These concentrations cover the normal range of most plasma amino acids. For plasma, the original plasma sample and plasma samples diluted by HPLC water to 75%, 50%, and 25% ($n = 4$) of the plasma were analyzed. The linear regression analysis showed satisfactory coefficients of correlation (>0.99) between the concentration and peak areas of each amino acid from both UV and fluorescent signals and in both standard amino acids and plasma samples (*11*).

References

1. Deyl, Z., Hyanek, J., and Horakova, M. (1986) Profiling of amino acids in body fluids and tissues by means of liquid chromatography. *J. Chromatog.* **379,** 177–250.
2. Qureshi, G. A. and Qureshi, A. R. (1989) Determination of free amino acids in biological samples: problems of quantitation. *J. Chromatog.* **491,** 281–289.

3. Williams, A. P. (1987) General problems associated with the analysis of amino acids by automated ion-exchange chromatography. *J. Chromatog.* **373,** 175–190.
4. Fekkes, D. (1996) State-of-the-art of high-performance liquid chromatographic analysis of amino acids in physiological samples. *J. Chromatog. B. Biomed. Appl.* **682,** 3–22.
5. Uhe, A. M., Collier, G. R., McLennan, E. A., Tucker, D. J and O'Dea, K. (1991) Quantitation of tryptophan and other plasma amino acids by automated pre-column o-phthaldialdehyde derivatization high-performance liquid chromatography: improved sample preparation. *J. Chromatog.* **564,** 81–91.
6. Sarwar, G. and Botting, H. G. (1993) Evaluation of liquid chromatographic analysis of nutritionally important amino acids in food and physiological samples. *J. Chromatog.* **615,** 1–22.
7. Furst, P., Pollack, L., Graser, T. A., Godel, H., and Stehle, P. (1990) Appraisal of four pre-column derivatization methods for the high-performance liquid chromatographic determination of free amino acids in biological materials. *J. Chromatog.* **499,** 557–569.
8. Worthen, H. G. and Liu, H. (1992) Automatic pre-column derivatization and reversed-phase high performance liquid chromatography of primary and secondary amino acids in plasma with photo-diode array and fluorescence detection. *J. Liq. Chromatogr.* **15,** 3323–3341.
9. Carducci, C., Birarelli, M., Leuzzi, V., Santagata, G., Serafini, P., and Antonozzi, I. (1996) Automated method for the measurement of amino acids in urine by high-performance liquid chromatography. *J. Chromatog. A.* **729,** 173–180.
10. Fekkes, D., van Dalen, A., Edelman, M., and Voskuilen, A. (1995) Validation of the determination of amino acids in plasma by high- performance liquid chromatography using automated pre-column derivatization with o-phthaldialdehyde. *J. Chromatog. B. Biomed. Appl.* **669,** 177–186.
11. Liu, H. and Worthen, H. G. (1992) Measurement of free amino acid levels in ultrafiltrates of blood plasma by high-performance liquid chromatography with automatic pre-column derivatization. *J. Chromatog.* **579,** 215–224.
12. Terrlink, T., van, L. P., and Houdijk, A. (1994) Plasma amino acids determined by liquid chromatography within 17 minutes. *Clin. Chem.* **40,** 245–249.
13. van Eijk, E. H., Rooyakkers, D. R., and Deutz, N. E. (1993) Rapid routine determination of amino acids in plasma by high- performance liquid chromatography with a 2–3 microns Spherisorb ODS II column. *J. Chromatog.* **620,** 143–148.
14. Georgi, G., Pietsch, C., and Sawatzki, G. (1993) High-performance liquid chromatographic determination of amino acids in protein hydrolysates and in plasma using automated pre-column derivatization with o-phthaldialdehyde/2-mercaptoethanol. *J. Chromatog.* **613,** 35–42.
15. Liu, H., Liu, Z. M., Zhu, W. N., Li, Y. H., Wang, S. D., and Jiang, X. S. (1984) Analysis of free amino acids in blood plasma by reversed phase high performance liquid chromatography with gradient elution and fluorescence detection. *Chinese J. Chromatog.* **1,** 83–87.

16. Ersser, R. S. and Davey, J. F. (1991) Liquid chromatographic analysis of amino acids in physiological fluids: recent advances. *Med. Lab. Sci.* **48,** 59–71.

17. Jones, B. N. and Gilligan, J. P. (1983) o-Phthaldialdehyde precolumn derivatization and reversed-phase high- performance liquid chromatography of polypeptide hydrolysates and physiological fluids. *J. Chromatog.* **266,** 471–482.

18. Schuster, R. (1988) Determination of amino acids in biological, pharmaceutical, plant and food samples by automated precolumn derivatization and high-performance liquid chromatography. *J. Chromatog.* **431,** 271–284.

19. Turnell, D. C. and Cooper, J. D. (1982) Rapid assay for amino acids in serum or urine by pre-column derivatization and reversed-phase liquid chromatography. *Clin. Chem.* **28,** 527–531.

20. Ogden, G. and Foldi, P. (1984) Amino acid analysis: An overview of current methods. *LC-GC* **5,** 28–40.

21. Lindroth, P. and Mopper, K. (1979) High performance liquid chromatographic determination of subpicomole amounts of amino acids by precolumn fluorescence derivatization with o-Phthaldialdehyde. *Anal. Chem.* **51,** 1667–1674.

22. Einarsson, S., Josefsson, B., and Lagerkvist, S. (1983) Determination of amino acids with 9-fluorenylmethyl chloroformate and reversed-phase high-performance liquid chromatography. *J. Chromatog.* **282,** 609–618.

23. Einarsson, S. (1985) Selective determination of secondary amino acids using precolumn derivatization with 9-fluorenylmethylchloroformate and reversed-phase high-performance liquid chromatography. *J. Chromatog.* **348,** 213–220.

24. Betner, I. and Foldi, P. (1988) The FMOC-ADAM approach to amino acid analysis. *LC-GC* **6,** 832–840.

25. Godel, H., Graser, T., Foldi, P., Pfaender, P., and Furst, P. (1984) Measurement of free amino acids in human biological fluids by high-performance liquid chromatography. *J. Chromatog.* **297,** 49–61.

26. Blundell, G. and Brydon, W. G. (1987) High performance liquid chromatography of plasma aminoacids using orthophthalaldehyde derivatisation. *Clin. Chim. Acta* **170,** 79–83.

27. Feste, A. S. (1992) Reversed-phase chromatography of phenylthiocarbamyl amino acid derivatives of physiological amino acids: an evaluation and a comparison with analysis by ion-exchange chromatography. *J. Chromatog.* **574,** 23–34.

28. Armstrong, M. D. and Stave, U. (1973) A study of plasma free amino acid levels. II. Normal values for children and adults. *Metabolism* **22,** 561–569.

29. Armstrong, M. D. and Stave, U. (1973) A study of plasma free amino acid levels. III. Variation during growth and aging. *Metabolism* **22,** 571–578.

11

Determination of Amino Acids in Foods by Reversed-Phase High-Performance Liquid Chromatography with New Precolumn Derivatives, Butylthiocarbamyl, and Benzylthiocarbamyl Derivatives Compared to the Phenylthiocarbamyl Derivative and Ion Exchange Chromatography

Kang-Lyung Woo

1. Introduction

Amino acid analysis with reverse-phase high-performance liquid chromatography (RP-HPLC) and ultraviolet (UV) detection following precolumn derivatization is popular owing to the greater versatility at the instrument, sensitivity and speed of analysis compared to specialized ion–exchange amino acid analyzers. Phenylthiocarbamyl (PTC) amino acid derivative is a precolumn derivatization method that has been widely used for analysis of amino acid by RP-HPLC *(1–8)*. This method is an excellent method for the derivatization of secondary amino acids, proline, and hydroxyproline *(2–4,7)*. However, its disadvantages are that it requires a high-vacuum system and it takes a long time to remove the byproducts produced in the process of derivatization and the excess reagent in order to avoid interfering peaks.

A need for simple, sensitive, stable, and more volatile precolumn derivatization reagent for analysis of amino acid with RP-HPLC and UV-detection still remains, although there have been developments of many reagents for precolumn derivatization, because these derivatives have some faults. The widely used precolumn derivatives, except PTC-amino acids, in RP-HPLC are *o*-phthalaldehydes (OPA) *(9,10)*, dansyl *(11,12)*, dabsyl *(13,14)*, and 4-nitrophenylthiocarbamyl (NPTC) *(7,15)* derivatives. With the OPA derivatives, secondary amino acids, proline, and hydroxyproline, were not detected

From: *Methods in Molecular Biology*, vol. 159: *Amino Acid Analysis Protocols*
Edited by: C. Cooper, N. Packer, and K. Williams © Humana Press Inc., Totowa, NJ

because OPA does not react with secondary amines in the absence of oxidizing agents. Moreover, as OPA derivatives are unstable, complete automation of the precolumn reaction with accurate control of the reaction time is essential for acceptable reproducibility (1).

Dansyl derivatives are formed in the dark and are unstable toward prolonged reaction times, solvents, and exposure to light and interfering peaks arise because of the byproducts during the derivatization process (12). In the dabsyl derivatives, it was reported that the sulphonamide bond in the derivative was very stable. The limitation of the dabsyl derivatization method is that the presence of an excess amount of urea, salt, phosphate, or ammonium hydrogen carbonate will change the pH of the reaction buffer and interfere with derivatization (13,14).

Derivatization with 4-nitrophenylisothiocyanate (NPITC) forms the stable nitrophenylthiocarbamyl (NPTC) derivatives that are suitable for analysis by RP-HPLC and UV-detection at 254 nm or 340 nm (15). The disadvantage of NPITC is that the excess reagent cannot be easily removed under a high-vacuum system. Extraction with toluene can remove the excess reagent (7).

A more volatile reagent for precolumn derivatization compared to other reagents developed to recent year, butylisothiocyanate (BITC), an aliphatic compound, was successfully adopted for derivatization to butylthiocarbamyl (BTC) derivatives of 22 protein standard amino acids. The BTC-amino acids were successfully analyzed via C_{18} RP-HPLC and UV-detector at 250 nm (16). The BITC reagent was also successfully adapted for the analysis of amino acids in foods (17).

The advantages of BITC were high volatility and the ability of separate derivatization on the cysteine and cystine, which had not been found with PTC-derivatives. The high volatility of this reagent substantially reduced the analysis time because the excess reagent and byproducts produced during the reaction could be easily removed. The BTC-derivatives of the secondary amino acids, proline, and hydroxyproline were also detected with high sensitivity. But asparagine and serine peaks overlapped completely and the stability of BTC-derivatives at room temperature was estimated to be only approx 8 h.

Benzylisothiocyanate (BZITC), the analog of phenylisothiocyanate (except NPITC), was successfully derivatized to benzylthiocarbamyl (BZTC) derivatives on all of the 22 protein amino acids and the derivatives were completely separated on a reversed-phase Nova-Pak C_{18} column (18). The BZITC reagent was less volatile compared to BITC but the volatility was similar to PITC. The advantages of BZTC-derivatives compared to PTC-derivatives was superior resolution on reversed-phase column and superior reproducibility to PTC-derivatives using the same experimental condition.

In this chapter, more detail experimental methods, results, and discussions on the BTC and BZTC-derivatives developed at the most recent years will be described.

2. Materials
2.1. Equipment

1. A water aspirator for vacuum (*see* **Note 1**).
2. Spectra-Physics 8800 ternary solvent delivery system with solvent stabilization and degassing system.
3. Spectra 200 programmable wavelength UV detector.
4. Nova-Pak C_{18} (300 × 3.9 id, 4 μm dimethyloctadecylsilyl-bonded amorphous silca, Waters).
5. Eppendorf CH-30 column heater.

2.2. Reagents

1. BITC, BZITC, and PITC are obtained from Aldrich (Milwaukee, WI) and were stored at 0–5°C (*see* **Note 2**).
2. Bovine serum albumin (BSA), standard amino acids, and norleucine are obtained from Sigma (St. Louis, MO). Standard amino acids were stored at room temperature and BSA was stored at 2–8°C.
3. HPLC-grade acetonitrile, methanol, and tetrahydrofuran from Merk (Darmstadt, Germany) were stored at room temperature.
4. All other reagents were of analytical grade.
5. Food samples, whole egg, and soybean purchased from a commercial market.

2.3. Solutions

1. Standard amino acid solution: A mixture solution of standard amino acids, except glutamine, cysteine, and cystine was prepared at a concentration of 2.5 μmol/mL of 0.01 M HCl. Standard solutions of glutamine and cysteine were prepared with water (*see* **Note 3**). Cystine was prepared at a concentration of 0.5 μmol/mL of 0.01 M HCl because of the solubility.
2. Preparation of coupling buffer solution: Coupling buffer (acetonitrile-methanol-triethylamine [10:5:2]) containing L-norleucine (2.5 μmol/mL) as an internal standard (*see* **Note 4**) was stored at 0–5°C and prepared again after 1 mo.

3. Method
3.1. Hydrolysis of BSA and Food Protein Samples (Figs. 1 and 2)

1. Place BSA (4 mg) and grinded soybean or homogenized whole egg food sample (0.2 g) in a 5-mL and 25-mL test tube, respectively, with an open-hole screw cap with a septum (**Fig. 3**).
2. Add 0.5 mL and 15 mL of 6 M HCl containing 0.1% phenol into the 5-mL and 25-mL test tubes containing the BSA and food samples, respectively.

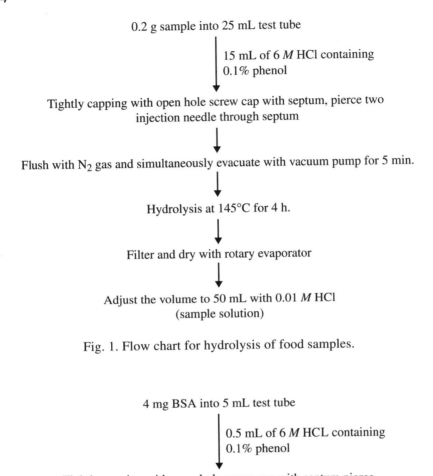

Fig. 1. Flow chart for hydrolysis of food samples.

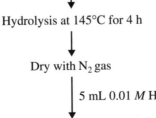

Fig. 2. Flow chart for hydrolysis of BSA.

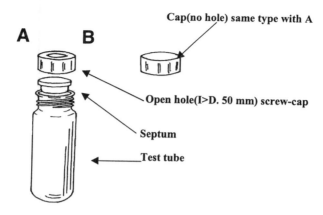

Fig. 3. Test tube for hydrolysis of food protein samples.

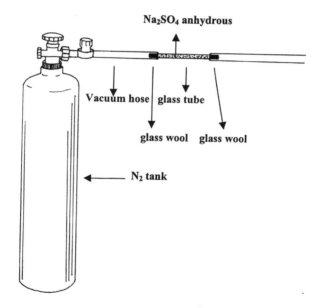

Fig. 4. The device for supply of the dried N_2 gas.

3. After tightly capping, pierce the septum with two stainless steel injection needles.
4. Connect one needle, immersed in the sample solution, to a dried nitrogen supply (*see* **Note 5**, **Fig. 4**). Connect the other needle, not immersed, to the vacuum pump (**Fig. 5**).
5. Evacuate the test tubes with a vacuum pump for 5 min and simultaneously flush with nitrogen gas.

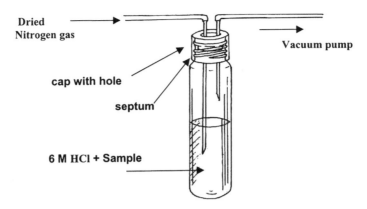

Fig. 5. The diagram of device for nitrogen saturation into the sample.

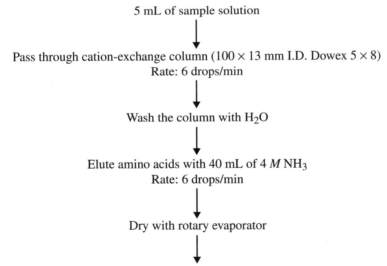

Fig. 6. Flow chart for clean up with cation–exchange column.

6. Remove the needle connected to the vacuum pump before removing the needle connected to the nitrogen.
7. Carefully remove the cap with the holes and change for a cap without holes (*see* **Note 6**), because small holes may be formed by the high pressure.
8. Carry out hydrolysis at 145°C for 4 h.
9. Dry the hydrolysate of BSA with nitrogen at 50°C. Redissolve with 5 mL of 0.01 M HCl (hydrolysate solution). This is now ready for derivatization.

10. Filter the hydrolysates of soybean meal and whole egg, then dry with a rotary evaporator. Redissolve and adjusted the volume to 50 mL with 0.01 M HCl (hydrolysate solution).
11. To clean up the hydrolysate solutions of soybean meal and whole egg use cation–exchange chromatography as follows (**Fig. 6.**).
12. Pass 5 mL of hydrolysate solution through a 100×13-mm id cation–exchange column (Dowex 5×8) at the rate of 6 drops/min, to retain the amino acids on the cation–exchange resin (*see* **Note 7**).
13. Wash the column several times with 20 mL of H_2O (*see* **Note 8**).
14. Elute the retained amino acids on the cation–exchange column with 40 mL of 4 M ammonia solution with the rate of 6 drops/min.
15. Dry the eluted solutions in a rotary evaporator at 50°C. Redissolve in 0.01 M HCl (hydrolysate solution) and adjust the volumes to 50 mL. These samples are now ready for derivatization (*see* **Note 9**).

3.2. Derivatization (Fig. 7)

1. Place 20 µL of the mixture solution of standard amino acids, 50 µL of the standard solution of cystine, and sample hydrolysates (BSA; 100 µL, soybean; 500 µL, whole egg; 500 µL) into separate 2-mL conical vials with an open-hole screw cap and a septum (**Fig. 8**).
2. Dry the solutions completely with nitrogen gas at 50°C.
3. Add an appropriate amount of acetonitrile to each vial and dried again (*see* **Note 10**).
4. Redissolve the residues in 50 µL of coupling buffer.
5. Add of 3 mL of BITC, BZITC, and PITC to each of the dissolved solutions for BTC, BZTC, and PTC derivatives, respectively.
6. After tightly capping the vials with open-hole screw-caps, the derivatizations are carried out at 40°C for 30 min for BTC and PTC derivatives and at 50°C for 30 min for BZTC derivatives.
7. After derivatizations, use two stainless steel injection needles to pierce through the septum into the vials. Connect one needle with the nitrogen supply and the other with the vacuum pump (**Fig. 9.**).
8. Infuse nitrogen into the vials and simultaneously evacuate with the vacuum pump to complete dryness at room temperature for approx 10 min for BTC derivative and for approx 40 min for BZTC and PTC derivatives.
9. Inject 100 µL of acetonitrile into the vials with a microinjection syringe and redry the contents for 5 min for BTC-derivative and for 30 min for BZTC and PTC derivatives (*see* **Note 11**).
10. Dissolve the residue of BTC-derivatives in 1 mL of 0.02 M ammonium acetate.
11. Dissolve the residue of BZTC and PTC-derivatives in 1 mL of 0.02 M NaH_2PO_4 containing 5% methanol and 1.5% tetrahydrofuran (pH 6.8, adjusted with phosphoric acid).
12. Filter the dissolved solutions through a 0.25-µm membrane filter.
13. Inject 10-µL aliquots of the filtrates onto the respective HPLC system.

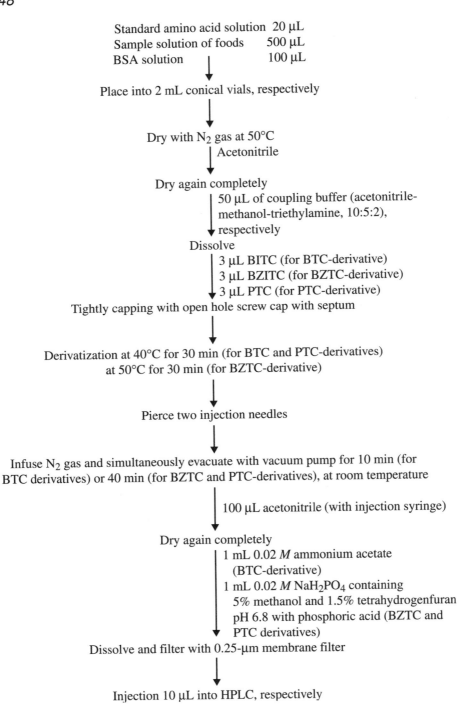

Fig. 7. Flow chart for derivatization.

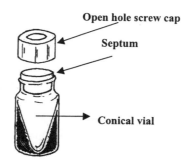

Fig. 8. 2-mL conical vial for derivatization.

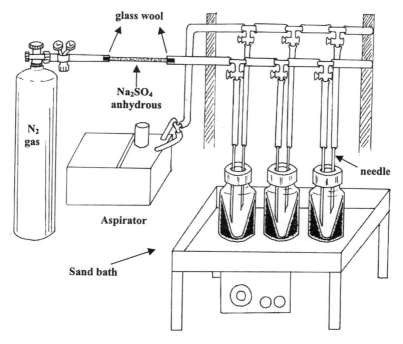

Fig. 9. The system drying several vials simultaneously.

3.3. Chromatography

HPLC conditions for the analysis were elucidated with **Table 1**.

3.4. Sensitivities of BTC, BZTC, and PTC-Derivatives

The sensitivities of BTC, BZTC, and PTC-derivatives are detected at 0.05 AUFS, which is the limit that gives a stable baseline with the smallest quantity.

It is possible to determine a linear relationship with the quantitative analysis (*see* **Notes 21** and **22**).

Table 1
HPLC Conditions for the Analysis of BTC, BZTC,
and PTC-Amino Acids Derivatives

HPLC system	Spectra-Physics 8800 ternary solvent delivery system
	Solvent stabilization and degassing system with a blanket of helium
Detector	Spectra 200 programmable wavelength UV detector
Wavelength for detection	240 nm for BTC-derivatives (*see* **Note 12**)
	246 nm for BZTC-derivatives (*see* **Note 13**)
	254 nm for PTC-derivatives (*see* **Note 13**)
Column	Nova-Pak C_{18} (300 × 3.9 id, 4 μm dimethyloctadecylsilyl-bonded amorphous silca, Waters). For all derivatives column-temperature; 40°C with Eppendorf CH-30 column heater
Solvent system	For BTC-derivatives (*see* **Notes 14–17**):

A solution; 0.05 M ammonium acetate (pH 6.7 adjusted with phosphoric acid)

B solution; 0.02 M sodium phosphate dibasic solution containing 5% methanol and 1.5% tetrahydrofuran-acetonitrile (50:50)

C solution; acetonitrile-water (70:30)

Solvent gradient

	A	B	C	Flow rate
0.0 min	100%	0%	0%	1 mL/min
5.0 min	85	15	0	"
14.0 min	70	20	10	"
20.0 min	60	20	20	"
25.0 min	30	20	50	"
30.0 min	10	20	70	"

For BZTC-derivatives (*see* **Notes 14–17**) and PTC-derivatives (*see* **Note 14–17**)

A solvent; 0.02 M NaH_2PO_4 containing 5% methanol and 1.5% tetrahydrofuran (pH 6.8 adjusted with phosphoric acid)

B solvent; A solvent-acetonitrile (50:50)

C solvent; acetonitrile-water (70:30)

Solvent gradient

	A	B	C	Flow rate
0.0 min	100%	0%	0%	1.2 mL/min
15.0 min	76	20	4	"
20.0 min	70	20	10	"
30.0 min	50	30	20	"
40.0 min	30	35	35	"

After this gradient program, a washing step for 20 min with solvent C substantially protected the column damage. Above two types of solvent gradients were adapted to the samples amino acid derivatives as well as standard amino acid derivatives (*see* **Note 18–20**).

3.5. Stabilities of the BTC-Derivatives and BZTC-Derivatives

The variations in the peak-area responses of BTC and BZTC-derivatives with storage time at room temperature can be determined for the stabilities of the BTC and BZTC-derivatives (*see* **Notes 23** and **26**).

3.6. Statistical Analysis

1. To determine the reproducibilities, all experiments should be repeated more than three times.
2. Relative standard deviations (RSDs) on the relative molar response (RMR) can be calculated for comparision of precisions on the derivatives. Linearity of calibration graphs on the appropriate ranges are also detected by the determination of statistical significance of correlation coefficient (γ) of calibration graphs (*see* **Notes 27** and **33**).
3. The reproducibility and accuracy on the BSA and food samples can be compared to ion–exchange chromatography and the data of the other literatures (*see* **Notes 34** and **38**).

4. Notes

1. If you use a vacuum pump that is not using water, you must install a device to absorb the evaporating excess reagents and byproducts produced during derivatization because the vacuum pump is rendered useless by these evaporating materials.
2. These reagents seem to be stable for several years unopened, but when the caps were opened several times, the reagents must be used in about 6 mo even if they are stored at –20°C. All of these reagents are very harmful and toxic.
3. Standard solutions of glutamine and cysteine were prepared with water because of the conversion to pyroglutamic acid and cystine, respectively, on prolonged storage of these amino acids in HCl solution *(19)*.
4. For BZTC-derivatives of sample internal standard (L-norleucine) should not be used.
5. For supply of the dried nitrogen, connect the tube for absorption of any moisture as shown in **Fig. 2**. This Na_2SO_4 anhydrous tube must be periodically changed every 1 or 2 mo.
6. When the cap with a hole is changed to the cap without a hole, you have to be carefully that the septum is not removed.
7. Be careful that the sample, washing, and eluting solutions do not drop under the level of cation–exchange resin in the column.
8. The rate of drops during the washing is not important.
9. For the analysis with ion–exchange chromatography and ninhydrin derivatization method, dry the clean-up sample solutions in a rotary evaporator. Redissolve in 0.2 *M* sodium citrate buffer (pH 2.2) and inject into the automation amino acid analyzer (LKB 4150 Alpha, Ultrapac-11 cation–exchange column).
10. By the synergy effect of acetonitrile evaporating, the residues of water are completely dried.

11. The chemical reactions of BTC and BZTC-derivatives are as follows:

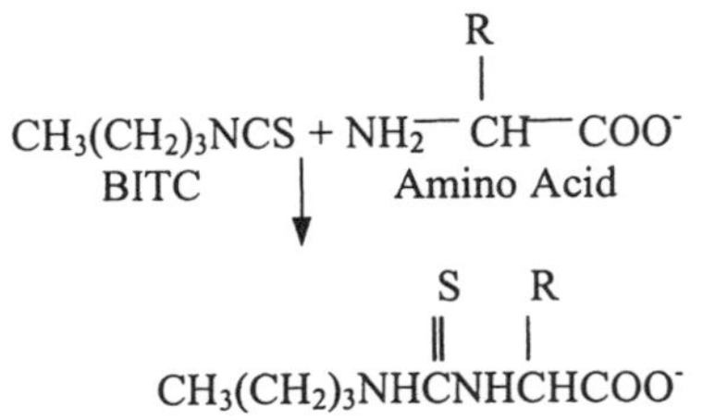

$$CH_3(CH_2)_3NCS + NH_2\overset{R}{\underset{|}{-}}CH-COO^-$$

BITC   Amino Acid

$$CH_3(CH_2)_3NH\overset{S}{\underset{\|}{C}}NH\overset{R}{\underset{|}{C}}HCHCOO^-$$

Butylthiocarbamyl(BTC) amino acid

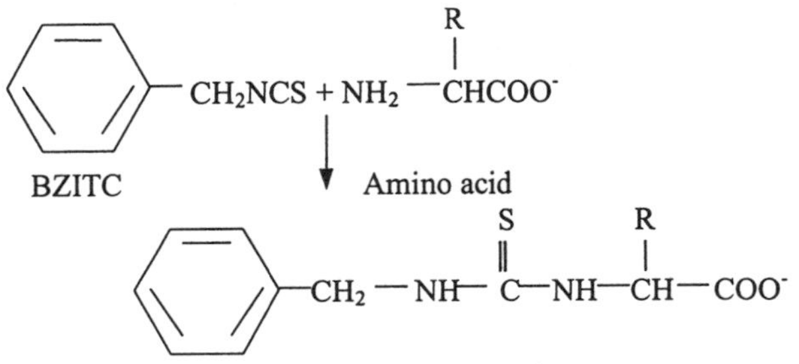

BZITC   Amino acid

Benzylthiocarbamyl(BZTC) amino acid

12. UV-spectra of the BTC-amino acid mixture and BITC, the coupling reagent, are shown in **Fig. 10**. The λ_{max} of BTC-amino acids was about 234 nm, but the most efficient wavelength was 250 nm, which avoided the absorption spectra of the impurities and the electrolyte, ammonium acetate in the solvent *(16)*.

13. The UV spectra of the BZTC-amino acids mixture and the PTC-amino acid mixture dissolved in 0.02 M NaH$_2$PO$_4$ are shown in **Fig. 11**. The wavelengths giving strong absorbance were 220 nm and 238 nm in the BZTC derivative and 215 and 270 nm in the PTC derivative, but the most efficient wavelengths were 246 nm in the BZTC derivative and 254 nm in the PTC derivative, which avoided interference by the absorption spectra of the impurities and electrolyte and showed a stable baseline *(18)*.

14. The standard amino acid chromatograms of BTC and BZTC-derivatives compared to PTC-derivatives separated on the Nova-Pak C$_{18}$ column are shown in **Fig. 12**. All of the 22 standard amino acids were derivatized with BTC, BZTC, and PTC-derivatives and resolved on C$_{18}$ reversed column.

15. In the BTC-derivatives, asparagine and serine completely overapped but BTC-cysteine and cystine were individually eluted even though the cystine peak was resolved with a tailing peak. In the PTC-derivatives, cystine and cysteine peaks appeared at the same position, from which it could be assumed that cysteine might be completely converted into cystine during derivertization because of the fact that cysteine can be oxidized *(20)*. Other articles have also reported that PTC-cysteine and cystine were eluted at the same position *(2,4,7)*.

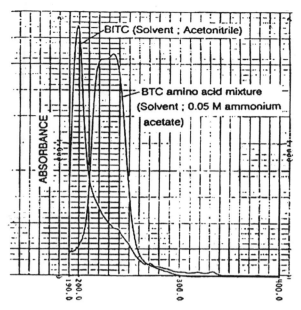

Fig. 10. UV spectra (*x*-axis in nm) of BTC-amino acid mixture and the coupling reagent BITC. Absorbance of solvents at 250 nm = 0.

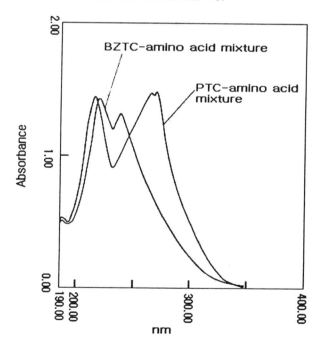

Fig. 11. UV spectra of the BZTC-amino acid mixture and the PTC-amino acid mixture. Solvent, 0.05 *M* NaH$_2$PO$_4$. Absorbance of solvent at 254 nm = 0.

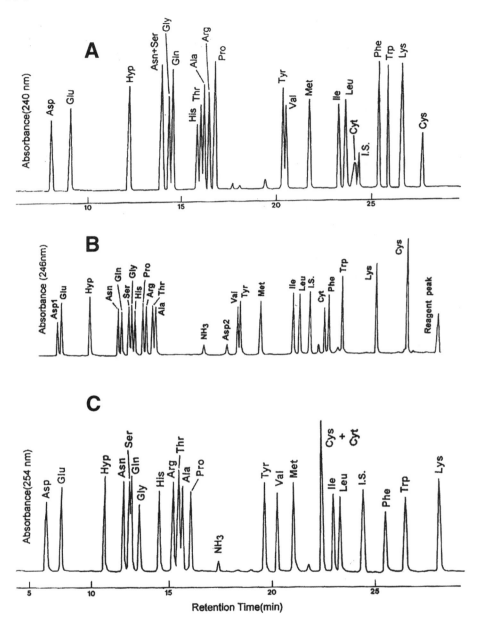

Fig. 12. Chromatogram of standard protein amino acid derivatives resolved on a
Nova-Pak (30 cm × 3.9 mm) C$_{18}$ column. I.S. = norleucine, injected amount 0.625
nmol. **(A)** BTC-amino acid; **(B)** BZTC-amino acid; **(C)** PTC-amino acid. Cyt = cystine; Cys = cysteine.

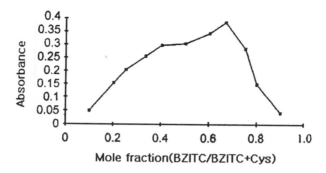

Fig. 13. Absorbance plot of BZTC-cysteine derivative at 238 nm showing maximum absorption of the mole fraction of BZITC in the derivatizing solution for the reaction between benzylisothiocyanate and cysteine.

16. Unlike BTC and PTC-derivatives, in the BZTC-derivatives all of 22 standard amino acids were nearly completely separated. Also, in the BZTC-derivatives as in the BTC-derivatives, the cysteine and cystine peaks were markedly separated.

17. Unlike the PTC-derivatives, in the BZTC and BTC-derivatives, the -SH moiety of cysteine would be converted to the thiocyanate derivatives by BZITC and BITC. To find out whether 1 mol of cysteine reacts with 2 mol of BZITC, absorbances of the BZTC-derivatives on the mole fraction of BZITC in the solution for derivertization between BZITC and cysteine were determined (**Fig. 13**). The determined wavelength was 238 nm, which gave maximum absorption for the BZTC-derivatives. Maximum absorbance appeared at 0.667 of the mole fraction, which meant that the mole ratio of the reaction between cysteine and BZITC was 1:2. So we could draw the conclusion that the -SH group, as well as the amino group of cysteine was derivatized. Intelligence due to the fact that -SH moiety of cysteine was readily converted to the thiocyanate derivative with 2-nitro-5-thiocyanobezoic acid *(21)* could support this conclusion. BZTC and BTC-cysteine might be eluted last because of the long less-polar side chain by the derivatization of –SH group.

18. Chromatograms of BTC, BZTC, and PTC-amino acid derivatives of soybean meal hyrolysate are shown in **Fig. 14**. In the BTC and PTC-derivatives we could detect the same kinds of amino acids of 16, but in the BZTC-derivatives cystine and cysteine in addition to above 16 amino acids were detected. We think that the sensitivity of BZTC-derivatives are more superior to the BTC and PTC-derivatives.

19. Chromatograms of BTC and PTC-amino acid derivatives on the whole egg hydrolysate are shown in **Fig. 15**. Because unfortunately, we did not detect on the BZTC-amino acid derivatives of the whole egg hydrolysate, it is really regretable that we could not compare to BTC and PTC-derivatives. The same kinds of 17 amino acids in the BTC and PTC-derivatives were detected. As is shown in the chromatograms of food samples (**Figs. 14** and **15**), assuming that the some kinds

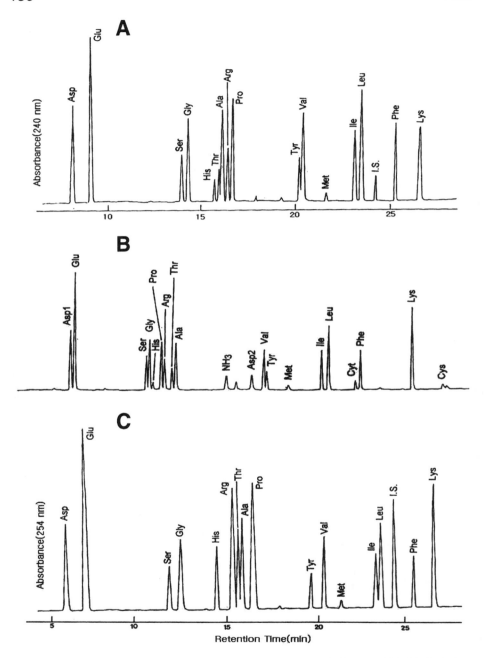

Fig. 14. Chromatogram of amino acids in soybean hydrolysate. (**A**) BTC derivatives; (**B**) BZTC derivatives; (**C**) PTC derivatives.

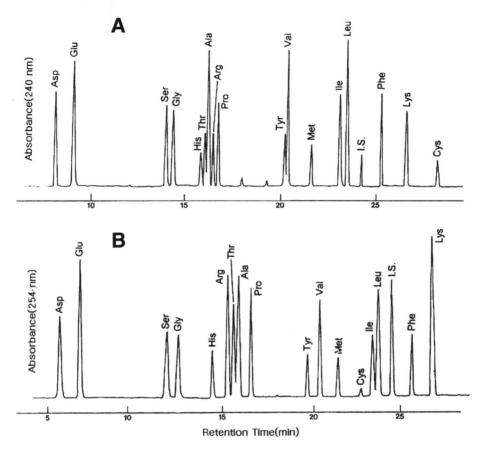

Fig. 15. Chromatogram of amino acids in whole egg hydrolyste. (**A**) BTC drivatives; (**B**) PTC derivatives. Cys = cysteine.

of contaminants were contained there were few ghost peaks. This phenomenon indicates that BITC, BZITC, and PITC are very good reagents having superior selective reactivity with amino acids.

20. Chromatograms at BTC, BZTC, and PTC-derivatives of BSA hydrolysate are shown in **Fig. 16**. In the soybean meal, on the other hand, the BTC and PTC-derivatives detected 17 amino acids, in the BZTC-derivatives 18 amino acids were detected with the additional detection of cystine.

21. Sensitivities of BTC, BZTC, and PTC-derivatives were about 3.9 pmol at 0.05 AUFS (**Fig. 17**). At levels lower than 3.9 pmol, several amino acids in the all of the three derivatives were not detected and not showed the linearity in the calibration graphs for the quantitative analysis.

22. There is a report that the sensitivity of the PTC-derivatives was 1 pmol in the detector at 0.005 AUFS (signal-to-noise ratio 5:1) (*2*). But in our experiment, at

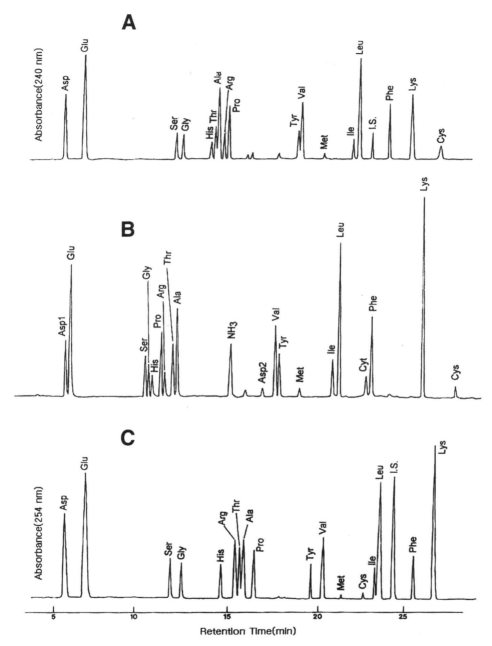

Fig. 16. Chromatogram of amino acids in bovine serum albumin hydrolysate. (A) BTC derivatives; (B) BZTC derivatives; (C) PTC derivatives. Cys = cysteine.

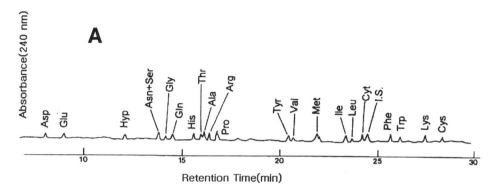

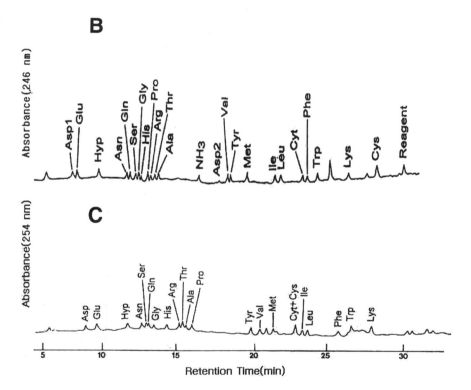

Fig. 17. Chromatogram of standard protein amino acid derivatives showing the sensitivity. Injected amount 3.9 pmol. Range 0.05 AUFS. (**A**) BTC-amino acid; (**B**) BZTC-amino acid; (**C**) PTC-amino acid. Cyt = cystine; Cys = cysteine.

this level, because the contamination of amino acids by the reagents, instrument, environment, and solvents was serious and the baseline was very unstable, the adaptation of the practical analysis was impossible.

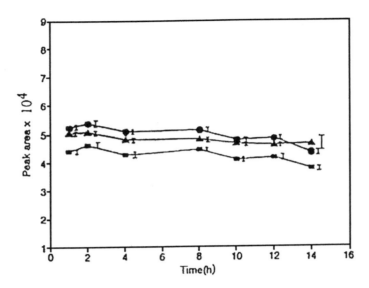

Fig. 18. Stability of BTC-amino acids at room temperature. ■ = Glu; ● = Ile; ▲ = Leu.

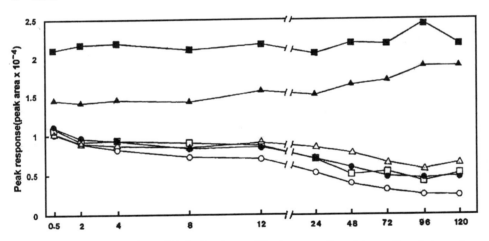

Fig. 19. Stability of BZTC derivatives of hydroxyproline (■), proline (▲), serine (●), cystine (□), alanine (△), and threonine (○). Each point is the mean value of three measurements.

23. The stability of BTC-derivatives and BZTC-derivatives is shown in **Figs. 18** and **19**, respectively.

24. In the BTC-derivatives, up to 2 h of storage, the peak area responses increased for most of the derivatives, which indicated the derivatization reaction was continuing at room temperature. After 8 h of the storage, the derivative that showed

Table 2
RMR of BTC, BZTC, and PTC Derivatives of Standard Amino Acids

Amino acid	BTC derivatives		BZTC derivatives		PTC derivatives	
	RMR[a]	RSD	RMR[b]	RSD	RMR[b]	RSD
Asp	0.50	1.40	1.89	3.82	0.89	2.48
Glu	0.48	1.25	1.21	0.83	0.84	1.83
Hyp	0.07	1.64	1.61	4.50	0.93	1.62
Asn	0.66[c]	0.45	1.12	4.48	0.91	2.37
Gln	0.71	1.27	1.07	2.80	1.03	5.89
Ser	—	—	1.05	1.46	0.76	2.02
Gly	0.52	0.96	0.96	2.63	0.65	1.68
His	0.62	1.13	0.84	2.49	0.83	0.87
Pro	0.71	1.55	1.17	3.42	0.85	7.37
Arg	0.44	0.68	1.00	1.15	0.91	3.02
Thr	0.29	1.72	0.92	1.09	1.01	16.6
Ala	0.75	2.53	0.88	1.73	0.90	6.64
Val	0.88	3.86	0.95	1.08	0.83	4.68
Tyr	0.83	1.33	1.39	2.20	0.92	2.64
Ile	0.84	0.03	1.05	0.95	0.85	3.76
Leu	0.86	2.21	0.84	1.38	0.84	6.86
Cyt	0.49	4.29	0.79	3.64	—	10.5
Phe	1.01	2.57	1.52	1.32	0.76	2.46
Trp	0.93	3.87	2.04	0.98	0.92	1.64
Lys	1.10	3.27	1.27	0.91	1.24	10.8
Cys	0.63	5.24	3.63	7.22	1.62[d]	—

RMR, relative molar response
RSD, relative standard deviation (n = 3)
[a]Values are relative to methionine.
[b]Values are relative to norleucine.
[c]Asn+Ser
[d]Cyt+Cys. Cyt = Cystine, Cys = cysteine.

the greatest decrease was cysteine (17.3%), the losses of the other derivatives after 8 h were in the range 0–6.9%. The derivatives that decreased to less than 5% after 14 h were glutamic acid, asparagine+serine, glutamine, threonine, tyrosine, proline, lysine, and tryptophan (*16*).

25. In the BZTC-derivatives, the peak area responses were decreased to less than 5% up to 120 h storage, but threonine, alanine, cystine, and serine were decreased to 18.9, 13.7, 15.1, and 15.6% after 4 h, respectively. Regarding the stability, we know that the optimum pH of the solvent dissolving the derivatives was very important. For the PTC-derivatives, the optimum pH of the solvent was 7.5. The loss of PTC-derivatives at this pH was in the range 0–10% after 10 h storage at room temperature (*7*).

Table 3
Amino Acid Compositions of Soybean Determined by the BTC, BZTC,
and PTC-Derivatives Compared to the Ion–Exchange Chromatography
(g/100 g dry matter)

Amino acid	BTC derivatives[a]	BZTC derivatives[b]	PTC derivatives[a]	PTC derivatives[b]	Ion-exchange chromatography[a]
Asp	4.17 ± 0.24	4.45 ± 0.25	4.64 ± 0.28	4.72 ± 0.63	4.25 ± 0.32
Glu	6.96 ± 0.42	7.41 ± 0.14	7.25 ± 0.34	7.76 ± 0.84	7.19 ± 0.12
Ser	1.61 ± 0.12	2.00 ± 0.16	1.59 ± 0.02	2.03 ± 0.43	1.65 ± 0.01
Gly	1.59 ± 0.07	1.56 ± 0.03	1.77 ± 0.10	2.09 ± 0.13	1.62 ± 0.05
His	1.50 ± 0.06	1.06 ± 0.08	1.52 ± 0.04	1.15 ± 0.04	1.32 ± 0.05
Arg	3.38 ± 0.17	3.61 ± 0.13	2.28 ± 0.08	2.65 ± 0.36	3.50 ± 0.09
Thr	1.19 ± 0.04	1.78 ± 0.11	1.79 ± 0.17	1.79 ± 0.05	1.24 ± 0.10
Ala	1.72 ± 0.10	2.00 ± 0.12	1.89 ± 0.35	1.89 ± 0.35	1.80 ± 0.12
Pro	2.15 ± 0.12	1.96 ± 0.20	2.06 ± 0.08	2.38 ± 0.18	2.20 ± 0.06
Tyr	1.20 ± 0.04	1.20 ± 0.05	1.29 ± 0.08	1.26 ± 0.11	1.15 ± 0.05
Val	2.88 ± 0.24	1.97 ± 0.10	2.47 ± 0.10	2.12 ± 0.10	3.01 ± 0.14
Met	0.42 ± 0.08	0.73 ± 0.01	0.54 ± 0.06	0.50 ± 0.25	0.40 ± 0.03
Ile	2.46 ± 0.14	1.70 ± 0.05	2.21 ± 0.06	1.83 ± 0.16	2.25 ± 0.18
Leu	3.38 ± 0.18	2.95 ± 0.18	3.03 ± 0.31	3.10 ± 0.24	3.57 ± 0.21
Cyt	—	0.49 ± 0.12	—	—	—
Phe	2.27 ± 0.16	2.06 ± 0.15	2.10 ± 0.10	2.63 ± 0.18	2.19 ± 0.14
Lys	3.03 ± 0.13	2.72 ± 0.10	3.15 ± 0.16	3.73 ± 0.25	2.86 ± 0.19
Cys	—	0.17 ± 0.04	—	0.11 ± 0.01[c]	—

All values are mean ± SD.
[a]$n = 4$. Data from **ref. 17**.
[b]$n = 5$. Data from **ref. 18**.
[c]Cys+Cyt. Cyt = cystine, Cys = cysteine.

26. We assumed that the stability of the BZTC-derivatives was superior to those of the PTC-derivatives and BTC-derivatives, except for threonine, alanine, cystine, and serine.
27. The relative molar responses of BTC, BZTC, and PTC-derivatives were shown in **Table 2**.
28. In the BTC-derivatives, RMR values determined with relative to methionine. RSDs on the RMR were less than 5% except cysteine (5.24%) in the BTC-derivatives. In the BZTC and PTC-derivatives, RMR values were determined with relative to norleucine, the internal standard.
29. In the BZTC-derivatives, the R.S.D values of all derivatives except cysteine (7.22%) were less than 5%. However, for the PTC-derivatives of glutamine, proline, alanine, and leucine the RSDs exceeded 5% and for threonine, cystine+cysteine, and lysine the RSDs exceeded 10%. In the PTC derivatization,

Table 4
Amino Acid Compositions of Whole Egg Determined by the BTC and PTC-Derivatives Compared to the Ion–Exchange Chromatography (g/100 g dry matter)

Amino acid	BTC derivatives	PTC derivatives	Ion-exchange chromatography
Asp	4.20 ± 0.09	4.74 ± 0.07	4.23 ± 0.05
Glu	5.48 ± 0.05	5.85 ± 0.59	5.76 ± 0.07
Ser	3.50 ± 0.09	2.61 ± 0.42	3.54 ± 0.10
Gly	1.56 ± 0.09	1.54 ± 0.31	1.53 ± 0.02
His	1.90 ± 0.14	1.91 ± 0.15	1.79 ± 0.06
Thr	1.75 ± 0.13	2.15 ± 0.11	1.82 ± 0.15
Ala	2.71 ± 0.16	3.29 ± 0.19	2.70 ± 0.11
Arg	3.55 ± 0.10	3.43 ± 0.48	3.61 ± 0.16
Pro	1.95 ± 0.15	1.87 ± 0.24	1.98 ± 0.22
Tyr	2.36 ± 0.62	1.96 ± 0.20	2.23 ± 0.30
Val	3.64 ± 0.20	3.83 ± 0.36	3.59 ± 0.32
Met	1.68 ± 0.11	1.58 ± 0.18	1.63 ± 0.06
Ile	3.42 ± 0.16	2.92 ± 0.28	3.26 ± 0.21
Leu	4.63 ± 0.21	4.32 ± 0.42	4.35 ± 0.09
Phe	2.94 ± 0.14	2.65 ± 0.22	2.88 ± 0.15
Lys	4.16 ± 0.20	4.30 ± 0.33	4.18 ± 0.30
Cys	1.35 ± 0.16	0.40 ± 0.10	1.41 ± 0.23

All values are mean ± SD, $n = 4$. Cys = cysteine.

the PITC reagent must be stored at –20°C under the inert gas to prevent breakdown *(7)* and a high-vacuum system (50–100 mtorr; 1 torr = 133.322 Pa) is needed to remove the excess reagent and byproducts that could interfere with the main peaks and to improve the reproducibility *(3)*. In this chapter, we did not use the high-vacuum system, so this seems to be one of the reasons why the RSD values in the PTC-derivatives exceeded 5%.

30. We concluded that BTC and BZTC derivatization was less fastidious compared to PTC derivatization.
31. Calibration graphs for BTC-derivatives showed good linearity in the range 0.5–2.5 nmol (not shown with Fig. or Table). The correlation coefficients (γ) of the calibration graphs for all of the derivatives were highly significant ($p < 0.001$), and the lowest value was $\gamma = 0.926$ for the cystine derivative, but it could be used for quantitative analysis.
32. Calibration graphs for all of the BZTC derivatives also showed good linearity in the measured range (0.125–5 nmol). The correlation coefficients (γ) of the calibration graphs were also highly significant ($p < 0.001$) and exceeded 0.99, except for cysteine (0.952) and glutamine (0.976).

Table 5
Amino Acid Compositions of BSA Determined by BTC and PTC-Derivatives Method Compared to the Ion–Exchange Chromatography and the Analytical Results Reported in the Literature

Amino acid	BTC derivatives[a]	BZTC derivatives[c]	PTC derivatives[a]	PTC derivatives[a,c]	Ion-exchange chromatography[a]	Automatic analyzer[d]	Sequence[e]	Sequence[f]
Asp	54.85 ± 3.01[b]	57.5 ± 4.76[b,g]	53.78 ± 4.75[b]	52.2 ± 0.42[b,h]	54.49 ± 3.51[b]	55.0	53	41
Glu	73.52 ± 5.04	79.8 ± 5.54	72.61 ± 0.94	78.5 ± 2.92	75.99 ± 4.14	82.0	78	59
Ser	24.60 ± 0.45	24.4 ± 1.60	26.22 ± 4.38	22.3 ± 1.92	25.84 ± 0.15	26.9	28	28
Gly	16.68 ± 0.28	13.8 ± 1.42	14.96 ± 0.50	16.3 ± 2.03	16.31 ± 0.37	15.3	15	16
His	22.39 ± 0.49	16.4 ± 3.22	20.85 ± 2.40	18.4 ± 0.70	17.60 ± 0.87	17.0	17	18
Arg	21.94 ± 0.22	23.0 ± 2.55	31.06 ± 1.51	23.9 ± 1.80	22.98 ± 0.17	23.6	23	25
Thr	30.35 ± 0.82	32.4 ± 1.52	31.16 ± 0.76	32.6 ± 0.80	32.05 ± 0.94	33.0	34	34
Ala	46.07 ± 0.94	47.6 ± 2.24	38.17 ± 0.27	40.0 ± 0.28	46.06 ± 1.33	45.2	46	46
Pro	26.85 ± 1.13	28.3 ± 2.65	29.27 ± 0.78	32.7 ± 0.35	28.60 ± 2.11	28.5	28	29
Tyr	20.47 ± 0.55	22.9 ± 1.56	20.72 ± 0.60	20.3 ± 2.12	20.22 ± 1.22	19.9	19	20
Val	38.61 ± 0.85	38.2 ± 1.50	36.87 ± 0.84	40.2 ± 2.42	37.28 ± 0.96	35.4	36	36
Met	4.61 ± 0.28	4.2 ± 0.25	6.97 ± 0.48	4.6 ± 0.56	4.43 ± 0.11	3.8	4	4
Ile	15.25 ± 0.29	15.9 ± 1.09	18.67 ± 0.55	16.9 ± 0.25	15.65 ± 0.32	14.1	14	14
Leu	61.77 ± 0.26	66.6 ± 2.02	60.88 ± 0.65	63.3 ± 3.68	61.13 ± 0.70	59.2	61	60
Phe	29.20 ± 1.07	28.7 ± 2.44	29.26 ± 0.65	30.3 ± 1.70	27.27 ± 0.80	2.1	26	26
Lys	58.75 ± 1.97	58.1 ± 2.73	54.58 ± 4.65	64.4 ± 2.42	59.42 ± 3.65	36.0	59	58
Trp	—	—	—	—	—	—	2	2
Cys	26.82 ± 3.60	12.4 ± 1.60	9.32 ± 0.59	25.8 ± 5.30[g]	28.66 ± 5.05	—	35	35
Cyt	—	12.5 ± 1.63	—	—	—	—	—	—
Gln	—	—	—	—	—	—	—	20
Asn	—	—	—	—	—	—	—	12
Total	572.73	582.7	555.35	582.7	573.98	582.2	578	583

Cys = cysteine.
[a] n = 3, Data from **ref. 17**.
[b] The values were recalculated on the basis of a total of 583 residues considering the fact that two Trp residues were completely destroyed and cysteine residues were substantially destroyed during acid hydrolysis.
[c] Data from **ref. 18**.
[d] Data from **ref. 24**.
[e] Data from **ref. 25**.
[f] Data from **ref. 26**.
[g] n = 12.
[h] n = 4.

33. In the PTC-derivatives, cystine+cysteine showed the lowest γ value (0.967), but it could be used for quantitative analysis. However, there is a report that the linearity for PTC-cystine was so poor that it could not be used for quantitative analysis *(22)*.

34. Amino acid compositions of soybean meal determined by the BTC, BZTC, and PTC-derivatives compared to the ion–exchange chromatography were shown in **Table 3**.

35. In the BTC-derivatives compared to ion–exchange chromatography, most of amino acids showed a deviation of less than 5%, except for histidine (13%) and isoleucine (9%). But BZTC and PTC-derivatives, most of amino acids showed a deviation of more than 5%. Reproducibility was superior in the BTC and BZTC-derivatives compared to PTC-derivatives.

36. **Table 4** shows the amino acid compositions of whole egg determined by the BTC and PTC-derivatives compared to ion–exchange chromatography. We did not determine BZTC-derivatives, so we could not show the data of the BZTC-derivatives. As in the soybean meal, the data of BTC-derivatives were well matched with those of ion–exchange chromatography. Reproducibility was also superior in the BTC-derivatives as in the soybean meal.

37. **Table 5** shows amino acid compositions of BSA recalculated on the basis of a total of 583 residues considering the fact that 2 Trp residues were completely destroyed, and cysteine residues were substantially destroyed during the hydrolysis.

38. It has been shown by sequence analysis that BSA has 17 disulfide bonds (cystine) and one free cysteine *(23)*. In this chapter, the number of BZTC-cystine and BZTC-cysteine residues was 12.5 and 12.4, respectively. It is known that a substantial amount of cysteine and cystine is destroyed during the hydrolysis of protein with 6 *M* HCl when the hydrolysis was carried out without conversion of these amino acids. Cysteine and cystine determination seems to give unavoidable error, unless they are modified to other compounds, i.e., pyridylethyl-cysteine or cystecic acid, prior to quantification *(18,21)*.

References

1. White, J. A. and Hart, R. T. (1992) Derivatization methods for liquid chromatographic separation of amino acid, in *Food Analysis by HPLC* (Nollet, L. M. L., ed.), Marcel Dekker, Inc. New York, pp. 53–74.

2. Bidingmeyer, B. A., Cohex, S. A., and Tarvin, T. L. (1984) Rapid analysis of amino acids using precolumn derivatization. *J. Chromatogr.* **336**, 93–104.

3. Heinrikson, R. L. and Meredith, S. C. (1984) Amino acid analysis by reverse-phase high-perfornance liquid chromatography; precolumn derivatization with phenylisothiocyanate. *Anal. Biochem.* **136**, 65–74

4. White, J. A., Hart, R. L., and Fry, L. C. (1986) An evaluation of the Waters Pico-tag system for the amino acid analysis of food materials. *J. Auto. Chem.* **8**, 170–177.

5. Beaver, R. W., Wilson, D. M., Johes, H. M., and Haydon, K. D. (1987) Amino acid analysis in foods and feedstuffs using precolumn phenylisothiocyanate derivatization and liquid chromatography-precolumn study. *J. Ass. Off. Anal. Chem.* **70**, 425–428.

6. Bidingmeyer, B. A., Cohen, S. A., Tarvin, T. L., and Frost, B. (1987) A new, rapid, high-sensitivity analysis of amino acids in food type samples. *J. Ass. Off. Anal. Chem.* **70,** 241–247.

7. Cohen, S. A. and Strydom, D. J. (1988) Amino acid analysis utilizing phenyliso-thiocyanate derivatives. *Anal. Biochem.* **174,** 1–16.

8. Koop. D. R., Morgan, E. T., Tarr, G. E., and Coon. M. J. (1982) Purification and characterization of a unique isozyme of cytochrome p-450 from live microsomes of ethanol-treated rabbits. *J. Biol. Chem.* **257,** 8472–8480.

9. Roth, M. (1971) Fluorescence reaction for amino acids. *Anal. Chem.* **43,** 880–882.

10. Yaegaki, K., Tonzetich, J., and Ng, A. S. K. (1986) Improved high-performance liquid chromatography method for quantitation of proline and hydroxyproline in biological materials. *J. Chromatog.* **356,** 163–???.

11. Tapuhi, Y., Schmidt, D. E., Lindner, W., and Karger, B. L. (1981) Dansylation of amino acids for high-performance liquid chromatography analysis. *Anal. Biochem.* **115,** 123–129.

12. DeJong, C., Hughes, G. J., Wieringen, E. V., and Wilson, K. J. (1982) Amino acid analysis by high-performance liquid chromatography. An evolution of usefulness of pre-column Dns derivatization. *J. Chromatogr.* **241,** 345–359.

13. Lin, J. K. and Chang, J. Y. (1975) Chromophoric labeling of amino acids with 4-dimethylaminoazobenzene-4'-sulfonyl chloride. *Anal. Chem.* **47,** 1634–1638.

14. Knecht, R. and Chang, J. Y. (1986) Liquid chromatograpic determination of amino acids after gas-phase hydrolysis and derivatization with (dimethylamino)azobenzene-sulfonyl chloride. *Anal. Chem.* **58,** 2375–2378.

15. Cohen, S. A. (1990) Analysis of amino acids by liquid chromatography after precolumn derivatization with 4-nitrophenylisothiocyanate. *J. Chromatogr.* **512,** 283–290.

16. Woo, K. L. and Lee, S. H. (1994) Determination of protein amino acids as butyl-thiocarbamyl derivatives by reversed-phase high-performance liquid chromatography with precolumn derivatization and UV detection. *J. Chromatogr. A.* **667,** 105–111.

17. Woo, K. L. Hwang, Q. C., and Kim, H. S. (1996) Determination of amino acids in the foods by reversed-phase high-preformance liquid chromatography with a new precolumn derivative, butylthiocarbamyl amino acid, compared to the conventional phenylthiocarbamyl derivatives and ion-exchange chromatography. *J. Chromatogr. A.* **740,** 31–40.

18. Woo, K. L. and Ahan, Y. K. (1996) Determination of protein amino acid as benzylthiocarbamyl derivatives compared with phenylthiocarbamyl derivatives by reversed-phase high-performance liquid chromatography, ultraviolet detection and precolumn derivatization. *J. Chromatogr. A.* **740,** 41–50.

19. Woo, K. L. and Lee, D. S. (1995) Capillary gas chromatographic determination of proteins and biological amino acids as N(o)-tert-butyldimethysilyl derivatives. *J. Chromatogr. B.* **665,** 15–25.

20. Greenstein, J. P. and Winitz, M. (1986) *Chemistry of Amino Acid*, vol. 3. Robert E. Krieger Publish., Malabar, FL, p. 1882.

21. Aitken, A., Geisow, M. J., Findlay, J. B. C., Holmes, C., and Yarwoord, A. (1989) *Peptide Preparation and Approach.* IRC, Oxford, pp. 43–68.
22. Fürst, P., Pollack, L., Graser, T. A., Gldel, H., and Stehle, J. (1990) Appraisal of precolumn derivatization methods for the high-performance liquid chromatographic determination of free amino acids in bilogical materials. *J. Chromatogr.* **499,** 559–569.
23. Dayhoff, M. O. (1976) *Atlas of Protein Sequence and Structure.* vol. 5, Suppl. 2, Natl. Biomed. Res. Found., Washington, DC, pp. 267.
24. King, T. P. and Spencer, M. (1970) Structural studies and organic ligand-binding properties of bovine plasma albumin. *J. Biol. Chem.* **245,** 6134–6148.
25. J. R. Brown (1975) Structure of bovine serum albumin. *Fed. Proc.* **34,** 591.
26. Hirayama, K., Akashi, S., Furuya, M., and Fukuhara, K. I. (1990) Rapid confirmation and revision of the primary structure of bovine serum albumin by ESIMS and Frit-FAB LC/MS. *Biochem. Biophys. Res. Commun.* **173,** 639–646.

12

Amino Acid Measurement in Body Fluids Using PITC Derivatives

Roy A. Sherwood

1. Introduction

Amino acid chromatography is used to detect both primary disorders of amino acid metabolism (e.g., maple syrup urine disease) and disorders of renal tubular reabsorption (e.g., cystinuria) (*see* **Notes 1–4**). In most patients with disorders in the former group, the abnormal amino acids are clearly increased in both plasma and urine, although the abnormality is usually more pronounced in urine. In disorders of renal tubular transport, only the urine amino acids will be abnormal, thus these conditions will be missed if plasma alone is studied. Quantitative amino acids are measured in blood samples from children who are having their amino acids levels manipulated by dietary restriction of natural protein. Derivatization of amino acids is required before analysis by high-performance liquid chromatography (HPLC). A number of alternative pre- or postcolumn derivatization methods have been described *(1)*. In the method described here amino acids in standards, urine, or deproteinized serum/plasma are reacted with phenylisothiocyanate (PITC) in the presence of a coupling solvent. The phenylthiocyanate (PTC) derivatives are then applied to a reverse-phase (RP) HPLC column and separated using a gradient elution system. The PTC amino acids are detected using an ultraviolet (UV) detector (254 nm) and an electrochemical detector (ECD) in series *(2)*.

2. Materials

2.1. Equipment

1. Gradient HPLC system with ODS Hypersil 5 μm, 4 mm × 25-cm column with precolumn filter.
2. UV detector and amperometric electrochemical detector.

From: *Methods in Molecular Biology*, vol. 159: *Amino Acid Analysis Protocols*
Edited by: C. Cooper, N. Packer, and K. Williams © Humana Press Inc., Totowa, NJ

Table 1
Composition of Coupling Reagent

Chemical	2 samples	10 samples	30 samples
Methanol (Hipersolv)	70 µL	350 µL	1050 µL
Triethylamine	10 µL	50 µL	150 µL
Distilled water	10 µL	50 µL	150 µL
neat PITC	10 µL	50 µL	150 µL

2.2. Reagents and Solutions

1. Acetic acid 6% (v/v).
2. Buffer A: 10 mM sodium acetate adjusted to pH 6.4 with 6% acetic acid. Should be prepared fresh for each run. Two liters will be enough to run 15 samples.
3. Buffer B: 10 mM acetonitrile, sodium acetate 60% (v/v) adjusted to pH 6.4 with 6% acetic acid. Stable at room temperature for 14 d. One liter will be enough to run 25 samples. Acetonitrile is toxic by ingestion, inhalation, and skin contact.
4. Precipitating reagent containing internal standard: 10% sulphosalicylic acid (SSA), 750 µM norleucine. Stable for 6 mo at 4°C.
5. PITC coupling reagent. Prepared fresh, as per **Table 1**. PITC is toxic by ingestion, irritating to skin/eyes, and has been reported to have teratogenic effects. Triethylamine is toxic by ingestion and is irritating to the eyes.
6. Amino acid standards (Sigma Chemicals, Poole, UK). Choice dependent on amino acids of interest. Available standards (17 amino acid standard, acids, and neutrals, basics) can be supplemented by specific amino acids as required.

3. Methods

3.1. Preparation of Samples

Plasma, random urine (children) or 24 h urine (adults) can be used (*see* **Notes 5** and **6**). Heparinized plasma is preferred, there are differences between plasma and serum amino acids. In acutely ill children with a suspected inherited metabolic defect, blood should be taken on admission. If the acute illness occurred some days before investigation, dietary protein may have been withdrawn and an amino acid abnormality may not be apparent. In a nonacute situation, a postabsorptive or fasting sample is preferred (adults and older children fasted overnight, smaller children 6–8 h). This is because plasma amino acids may be significantly increased 1–2 h after a protein meal. In neonates samples should be taken 2–3 h after feeding.

1. Urine. For urine, determine the creatinine in mmol/L (*see* **Note 7**). For creatinine concentrations below 5 mmol/L, take 50 µL of urine, make up to 100 µL with distilled water, and multiply result by two. For urine creatinine concentrations

above 5 mmol/L, take 20 μL of urine, make up to 100 μL with distilled water, and multiply result by 5. For urine creatinine concentrations below 1 mmol/L, another sample should be requested as the urine is too dilute to give meaningful results.

2. Add 50 μL of SSA precipitating reagent to tube. Mix and leave for 30 min at 4°C.
3. Plasma/serum: Pipet 100 μL of sample into a microfuge tube. Add 50 μL of SSA precipitating reagent to tube. Mix and leave for 30 min at 4°C.
4. Standards: 10 μL added to 90 μL of distilled water. Additional amino acid standards can be added at this point with corresponding reduction in the water, e.g., add 10 μL of a 1 mg/mL standard. Standards as prepared give 250 μmol/L concentrations except cystine (125 μmol/L). Add 50 μL of precipitating reagent containing internal standard to tube. Mix and leave for 30 min at 4°C.

3.2. Derivative Formation

1. Centrifuge tubes for 10 min at 14,000g.
2. Transfer 50 μL of supernatant to an appropriately labeled microfuge tube. Add 200 μL acetonitrile and mix.
3. Add 40 μL PITC/coupling reagent and vortex mix. Leave to react for 20 min.
4. Evaporate to dryness at a temperature less than 45°C. *It is essential that the derivatives are dry, but not cooked/desiccated.* The derivatives (dry) can be left at this stage at 4°C (no more than 72 h).
5. Dissolve in 200 μL of sodium acetate buffer (Buffer 1). Mix.
6. Centrifuge and remove 200 μL of supernatant to HPLC vial with microinsert for chromatography.

3.3. Chromatography

1. HPLC operating conditions: Column oven temperature: 30°C, sample cooler, 4°C, injection volume 25 μL. Electrochemical detector settings: potential +1.1 V (oxidation), sensitivity 3 μA. UV detector: wavelength 254 nm. If the UV detector and the electrochemical detector are both used in series the UV detector must be placed first.
2. Gradient: The solvent gradient is set as shown in **Table 2**.
3. Allow the system to equilibrate and run a gradient off-line while preparing samples and standards. The cycle time on the autosampler is 95 min.
4. Inject the first standard in the queue × 2. (The first sample retention times are often too variable for calibration purposes.)
5. Typical retention times and relative (to the internal standard norleucine) retention times are shown in **Table 3** (*see* **Notes 8–12**).
6. Calculation: To obtain calculation factor:

$$\text{Calculation factor} = \frac{\text{conc of std (μmol/L)} \times \text{area/height of Internal Standard}}{\text{area/height of Standard}}$$

The peak area/height of unknowns is multiplied by this factor to give the concentration in μmol/L.

Table 2
HPLC Solvent Gradient

Time (min)	Buffer 1 (%) Sodium acetate, pH 6.4	Buffer 2 (%) 60% acetonitrile/ sodium acetate, pH 6.4	Flow rate mL/min
0	100	0	1.0
20	87	13	1.0
65	45	55	1.0
67.5	0	100	2.0
70	0	100	2.0
75	100	0	2.0
80	100	0	1.0
85	100	0	1.0
95	100	0	1.0

7. Interfering substances: Vigabatrin will form a PTC derivative. The cephalosporin group of antibiotics potentially interfere with the UV detection method and might cochromatograph with glycine.

4. Notes

1. In approx 99% of cases, the amino acid pattern will be within normal limits. Variations occur for physiological reasons, particularly in premature neonates, but in comparison to pathological processes the variation is small. Most inherited disorders of amino acid metabolism produce gross changes in the observed pattern. Children diagnosed with inherited diseases involving amino acids are often treated with artificial feeds excluding certain amino acids, and they can get very low levels of essential amino acids, necessitating frequent monitoring.

2. Pathological cases: Most are fairly obvious and the specific amino acid will increase dramatically, e.g., in phenylketonuria, cystinuria, maple syrup urine disease, cystathioninuria, and hypophosphatasia the increase on HPLC will be off scale by 2–10 fold. Homocystinuria is an exception to this in that homocystine is normally undetectable by HPLC. A positive cystine screen with a normal cystine on thin layer chromatography should alert you to the possibility of homocystinuria. Using the HPLC method, it chromatographs immediately after the internal standard and any visible peak must be considered abnormal, but this may not be easy to pick up. Methionine concentrations are often high, secondary to liver disease, such as tyrosinemia and galactosemia.

3. Symptoms suggestive of specific disorders:
   a. Cataracts. Common in homocystinuria.
   b. Deafness. Occurs in arginosuccinic aciduria. Gives three peaks on HPLC (two anhydrides and the acid itself).

Table 3
Mean Retention Times and Relative Retention Times for the Amino Acids Most Likely to be Found in Body Fluids

Amino acid	Retention time (min)	Relative retention time
Phosphoserine	3.82	0.084
Aspartic acid	5.44	0.12
Glutamic acid	6.83	0.15
γ-Aminoadipic acid	10.35	0.229
Hydroxyproline	12.25	0.271
Phosphoethanolamine	12.45	0.277
Serine	14.52	0.321
Glycine	15.53	0.345
Asparagine	15.55	0.347
Sarcosine	16.91	0.375
β-Alanine	17.40	0.387
Taurine	19.00	0.421
γ-Aminobutyric acid	20.47	0.454
Citrulline	20.65	0.458
Threonine	21.11	0.468
Alanine	21.75	0.482
β-Aminoisobutyric acid	22.27	0.494
Proline	23.49	0.521
Histidine	23.91	0.532
Carnosine	25.71	0.569
Arginine	28.22	0.626
Methyl histidine	28.39	0.630
α-Aminobutyric acid	28.60	0.634
Anserine	29.06	0.644
Tyrosine	35.02	0.778
Valine	35.76	0.795
Ethanolamine	37.66	0.835
Methionine	38.10	0.845
Cystathionine	38.20	0.848
Cystine	40.88	0.907
Isoleucine	42.79	0.950
Leucine	43.47	0.965
Norleucine	45.00	—
Hydroxylysine	47.78	1.060
Phenylalanine	48.20	1.069
Ornithine	49.67	1.102
Tryptophan	50.16	1.112
Lysine	52.79	1.170

    c. Abnormal X-ray. Hypophosphatasia (serum alkaline phosphatase activity will be < 25 IU/L).

    d. Renal Stones. Cystinuria (also possibly xanthine/purine stones which are radiolucent in disorders such as Lesch-Nyhan syndrome).

    e. Organic acidurias. Most organic acidurias produce a metabolic acidosis, a urine pH < 6.0 and coma. Some include hypoglycemia and abnormal blood lactate/pyruvate ratio. There are often abnormalities in the amino acid pattern, but these are not always helpful.

4. Physiological/spurious increases in amino acids:

    a. Tyrosine. Transient neonatal tyrosinaemia is common in the first 6–8 wk of life for premature infants. If it persists after this time, then either tyrosinaemia or liver disease should be suspected (the latter is about 10 times more likely). The cutoff for tyrosine levels giving rise to recall in neonatal screening programs is typically 500 μmol/L. However, recent experience has shown that tyrosine concentrations in tyrosinemia type I can be around 200 μmol/L with a neonatal presentation.

    b. Threonine. Increased in neonates when compared to adults. It is also increased in liver disease.

    c. Taurine. Increased taurine can occur in catabolic states, but there is no specific disorder involving taurine.

    d. Glycine/aspartic acid/glutamic acid. Hyperglycinemia can occur in organic acidurias, valproate treatment, and in subjects with a low calorie intake. In urine samples with bacterial contamination, glutamine is deaminated to glutamic acid, aspartic acid may be increased, and hippuric acid is broken down to glycine. A pH >7.0 is suspicious in such cases.

    e. Lysine/cystine. Both are often raised in the first 6 mo of life.

       A generalized amino aciduria is often seen in very sick neonates with poor peripheral circulation. Particular increases are seen in alanine, glycine and proline related to disturbances of lactate/pyruvate metabolism.

5. Samples should *not* be hemolyzed. Samples are stored *frozen* if not assayed immediately. Samples left unseparated will have increases in taurine, phosphoethanolamine, aspartic, and glutamic acid. Free cystine and homocystine in plasma should only be measured in samples that have been separated and frozen within 30 min of being taken. Care should be taken not to aspirate the buffy coat with the plasma, this leads to increases in taurine, aspartic acid, glutamic acid, and phosphoethanolamine.

6. A 24-h collection is required for monitoring treatment of cystinuric patients (a compromise for adults is to fast overnight and collect a 4-h (still fasting) collection (from 0630–1030) *(3)*.

7. An aliquot of urine should be taken on receipt for creatinine estimation, the remainder of the urine should be frozen until analyzed. Excretion rates of amino acids may vary independently of the creatinine over a 24-h period. Diurnal variation has been observed.

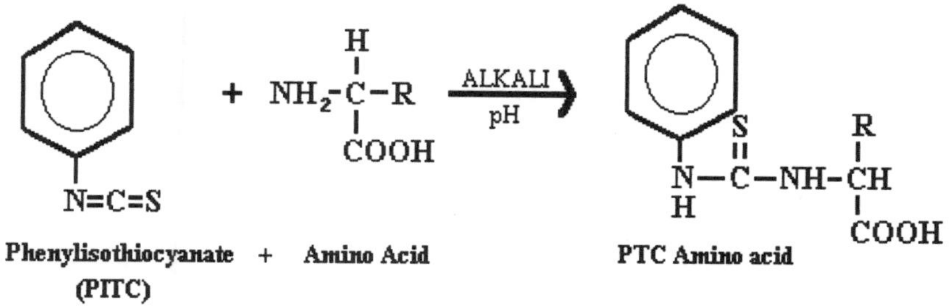

Fig. 1. Formation of the PTC derivatives of amino acids.

8. Identify peaks on samples by cross reference to standards and to relative retention times obtained on previous runs. Peaks can be identified by cochromatography. Mix equal volumes of a derivatized standard and sample, cochromatography and the amino acid is probably the same as in the sample, splitting of the peak — it is not the same.

9. Several of the peaks are pH dependent and move slightly as the buffer ages, in addition, separations tend to vary with different batches of the column. Typically, proline moves to merge with histidine or alanine. Hydroxyproline, ethanolamine, and phosphoethanolamine also move, but this is seldom a problem.

11. Histidine can be identified by its characteristic tailing peak.

12. Threonine and citrulline tend to cochromatograph, citrulline runs slightly fast of threonine.

13. Glycine and asparagine occasionally run together, glycine is faster than asparagine.

Acknowledgments

The sulphosalicylic acid precipitation modification of the original method was a personal communication by Dr. Steve Kryawych (Great Ormond Street Hospital).

References

1. Walker, V. and Mills, G. A. (1995) Quantitative method for amino acid analysis in biological fluids. *Ann. Clin. Biochem.* **32,** 28–57.

2. Sherwood, R. A., Titheradge, A. C., and Richards, D. A. (1990) Measurement of plasma and urine amino acids by high-performance liquid chromatography with electrochemical detection using phenylisothiocyanate derivatization. *J. Chromatog.* **528,** 293–303.

3. Parvy, P. R., Bardet, J. I., Rabier, D. M., and Kamoun, P. (1988) Age related reference ranges for free amino acids in first morning urine specimens. *Clin. Chem.* **34,** 2092–2095.

13

Determination of Proteins, Phosphatidylethanolamine, and Phosphatidylserine in Lipid-Rich Materials by Analysis of Phenylthiocarbamyl Derivatives

Margareta Stark and Jan Johansson

1. Introduction

In classical, ninhydrin-based amino acid analysis *(1,2)*, the ion–exchange matrix used for separation becomes contaminated upon consecutive analyzes of extremely lipid-rich samples; in our experience, already after approx 20–30 samples. Therefore, in analysis of lipid-rich material we focused on amino acid analysis involving reversed phase (RP) chromatography, because lipids are soluble in the organic solvents commonly used for elution and regeneration of such columns (e.g., acetonitrile and 2-propanol). Precolumn derivatization with phenylisothiocyanate (PITC) of protein hydrolysates from physiological samples followed by RP-HPLC is equivalent in analytical quality to the classical ion–exchange chromatography/ninhydrin method *(3)*, and is sensitive down to at least 10 pmol *(4)*. PITC reacts both with primary and secondary amines, the reproducibility is high and the phenylthiocarbamyl (PTC) amino acid derivatives are stable for months when stored dry at –20°C, or for days in solution at ambient temperature *(3,4)*. This approach can be used conveniently for analysis of lipid-rich material, where at least 300 samples can be analyzed with the same column, i.e., the column lifetime is comparable to those encountered during analysis of lipid-free samples *(5)*.

During isolation of proteins from lipid-rich sources, we have found it necessary to monitor both protein and phospholipid profiles, at least at early stages of the purification scheme. The phospholipids are usually detected by phosphorous analysis, but with the PTC/RP-HPLC method, phospholipids contain-

From: *Methods in Molecular Biology*, vol. 159: *Amino Acid Analysis Protocols*
Edited by: C. Cooper, N. Packer, and K. Williams © Humana Press Inc., Totowa, NJ

ing a primary amino group can be determined simultaneously with the proteins. Phosphatidylethanolamine subjected to hydrolysis and subsequent analyzed by the PTC/RP-HPLC method show a peak eluting between PTC-Arg and PTC-Tyr, with a linear range between 1–50 nmol, but lower levels are detectable. With the classical ion–exchange/ninhydrin amino acid analysis, where ethanolamine coelutes with Lys and consequently neither of these compounds can be determined. Phosphatidylserine analyzed in the same way by the PTC/RP-HPLC method gives one peak that coelutes with PTC-Ser, with a linear range between 10 pmol–50 nmol. When monitoring the protein and phospholipid profile after chromatography, it is feasible to determine when serine derived from phosphatidylserine elutes by looking at the overall PTC-amino acid profile and amounts. Phosphatidylcholine contains a quaternary amino group and cannot be detected with the PTC/RP-HPLC method *(5)*. A major advantage with PTC/RP-HPLC analysis, in addition to allowing repeated analysis of lipid-rich samples, is that phosphatidylethanolamine and phosphatidylserine can be determined simultaneously with the proteins, thereby reducing the need for separate phosphorous analysis during protein purification. A sample from a lung phospholipid fraction, which contains about 2% protein and 98% lipids, was hydrolyzed and analyzed by the PTC/RP-HPLC method (**Fig. 1**).

2. Materials

2.1. Hydrolysis

1. Glass tubes for hydrolysis and derivatization are 6–7 × 35 mm. Tubes for mixing derivatization solutions are 10 × 75 mm. Submit all glass tubes to pyrolysis (400–500°C for 3–4 h) before use to remove any contaminating material. Recommended glass quality: soda-lime or Duran.
2. Hydrolysis solution: 6 *M* HCl with 0.5% (w/v) phenol.

2.2. Derivatization

1. Amino acid standard in 0.1 *M* HCl (e.g., Amino Acid Standard H, Pierce). A stock solution of 100 μ*M* of each amino acid in water can be stored at –20°C for several months.
2. Phosphatidylethanolamine (e.g., dipalmitoylphosphatidylethanolamine, Fluka) dissolved in chloroform, e.g., 10 m*M* stock solution. Store in capped glass tube at –20°C.
2. Triethylamine (sequanal grade, Pierce). Store at 4°C. NB! Avoid inhalation and skin contact. Wear protective clothing. Work in a ventilated hood and use gloves.
3. Phenylisothiocyanate (PITC) (sequanal grade, Pierce). Store opened ampoule in a screw-capped glass tube under nitrogen in the dark at ambient temperature. Discard when yellow. Caution: Avoid inhalation and contact with skin and eyes. Work in a ventilated hood, use gloves, and wear protective clothing.

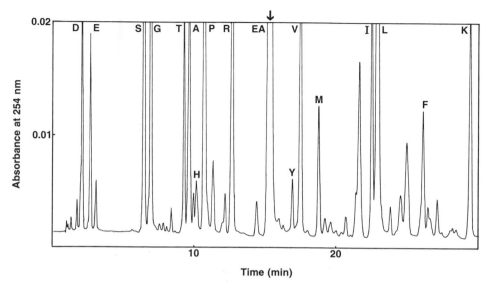

Fig. 1. RP-HPLC of PTC derivatives from a sample obtained from a phospholipid/ Lipoprotein fraction from porcine lung. The sample was hydrolyzed and analyzed by the PTC/RP-HPLC method. The PTC derivatives are identified with the one-letter code for the corresponding amino acid or EA for ethanolamine. The arrow identifies the peak that appears when dipalmitoyl-phosphatidylethanolamine was hydrolyzed separately.

4. Pre-derivatization solution: 99.5% ethanol/water/triethylamine (2:2:1, v/v/v), 40 μL per sample is required. Mix in a 10 × 75-mm glass tube. Prepare fresh.
5. 50% ethanol: 99.5% ethanol/Milli-Q water (1:1, v/v). Store in a screw-capped glass tube at 4°C.
6. Derivatization solution: 99.5% ethanol/triethylamine/PITC (7:2:1, v/v/v). Mix in a 10 × 75-mm glass tube. Prepare fresh.

2.3. Separation

1. Mobile phase A: 30 mM phosphate buffer, pH 6.30–6.80 (*see* **Note 1**). Dissolve 1.2 g NaOH in approx 800 mL MilliQ water, adjust to the pH required with 1 M H_3PO_4. Adjust to 1000 mL and filter through 0.2 μm. Degas before use. Use fresh or store at 4°C overnight. If stored longer (a few days), filter again through 0.2 μm before use.
2. Mobile phase B: 60% acetonitrile in water. Mix 600 mL acetonitrile with 400 mL Milli-Q water. Store at ambient temperature. Stable for several weeks. Degas before use.
3. Column: C_{18} reversed phase HPLC column, Spherisorb S3 ODS2, 4.6 × 100 mm, packed with 3-μm particles (Waters).

Table 1
Gradient Program for Separation of the PTC-Derivatives

Time (min)	% A	% B
0	100	0
1	95	5
26 (**Note 6**)	68 (66)	32 (34)
28	0	100
34	0	100
37	100	0

Flow rate 1.0 mL/min, the gradient is linear. Next sample injected after 8 min delay (45 min total per sample).

4. High-performance liquid chromatography (HPLC) instrumentation: Any analytical instrument equipped with two separate pumps, absorbance detector (254 nm), integrator, heating block, and a temperature control unit.

3. Methods

3.1. Hydrolysis

1. Pipet samples/phospholipid standards (*see* **Note 2**) present in volatile solution in 6 × 35-mm glass tubes and dry the samples under a stream of nitrogen or vacuum. For the acid hydrolysis add 40 µL 6 *M* HCl/0.5% phenol to each tube, evacuate and seal the tubes. Incubate at 110°C for 20–24 h, open and dry the samples under vacuum.

3.2. Derivatization

1. Pipet amino acid standard (1000–5000 nmol/aa) in a 6 × 35-mm glass tube and dry under vacuum (*see* **Note 3**).
2. Add 40 µL pre-derivatization solution (99.5% ethanol/water/triethylamine; 2:2:1) *(6)* to each of the hydrolyzed samples/phospholipid standards and amino acid standard, seal the tubes with parafilm, and mix for 10 s. Remove the parafilm and redry under vacuum (*see* **Note 4**).
3. For derivatization add 3 µL 50% ethanol to each tube *(7)*, make sure that an aqueous film is obtained at the bottom of the tube by mixing or turning the tube. Then add 7 µL of derivatization solution (99.5% ethanol/triethylamine/PITC, 7:2:1) and seal the tubes with parafilm. Mix vigorously (vortex) for 10 s and allow the reaction to occur for 15–30 min at ambient temperature. Remove the parafilm and dry the samples under vacuum overnight to remove excess reagent (*see* **Note 5**).
4. Seal the tubes with parafilm and store at –20°C, if not immediately subjected to chromatography. The samples can be stored for several months in the freezer.

3.3. Separation of PTC-derivatives by HPLC

1. Dissolve the derivatized and dried standard and samples in 100 µL mobile phase A. Filter lipid-containing samples through 0.45 µm.
2. Equilibrate the column with mobile phase A, flow rate 1.0 mL/min, and column temperature 36°C.
3. Inject 5–90 µL standard or sample on the column and carry out gradient elution as shown in **Table 1** at a flow rate of 1.0 mL/min and a column temperature of 36°C. Allow the column to equilibrate with mobile phase A for 8 min between each run, i.e., one sample injected every 45 min.
4. Quantify the PTC-amino acids and PTC-ethanolamine by comparison of the peak areas with those of the standard mixtures.
5. Store the column between runs in 60% acetonitrile (*see* **Note 7**).

4. Notes

1. Start at pH 6.3–6.4. The elution position of His is pH sensitive. Lower pH — earlier retention time of His, higher pH — later retention time. After several runs His usually elutes close to, or coelutes with Pro, then raise the pH (0.05–0.1 pH U). Do not exceed pH 7.0, the matrix of the column is not stable in basic pH.
2. Phosphatidylethanolamine standard, suggested amounts: hydrolyze 10 nmol and analyze 10% (i.e., 1 nmol) or make a dilution series. Analysis of less than 500 pmol gives unreliable results. A phosphatidylserine standard is not necessary, the serine from the amino acid standard can be used.
3. Do not hydrolyze the amino acid standard. 50 pmol per amino acid analyzed is suitable for AUFS 0.005 at 254 nm.
4. This step is indispensable. PITC do not react with the amino groups if any residual acid from the hydrolysis is present.
5. PITC has low volatility. Remove by high vacuum in a dessicator with solid NaOH for at least 12 h. If the reagent peak is to high, redry for longer time (e.g., 24 h).
6. Lower %B — later retention times from Ser. Higher % B — earlier retention times from Ser.
7. After 30–50 runs, wash the column with methanol/2-propanol, 1/1 (v/v) to minimize lipid contamination. Store the column in at least 60% acetonitrile.

Acknowledgments

We are grateful to Carina Palmberg for skillful assistance and Professor Hans Jörnvall for support.

References

1. Spackman, D. H., Stein, W. H., and Moore, S. (1958) Automatic recording apparatus for use in chromatography of amino acids. *Anal. Chem.* **30**, 1190–1206.
2. Hamilton, P. B. (1963) Ion exchange chromatography of amino acids. A single column, high resolving, fully automatic procedure. *Anal. Chem.* **35**, 2055–2064.

3. Sarwar, G. and Botting, H. G. (1993) Evaluation of liquid chromatographic analysis of nutritionally important amino acids in food and physiological samples. *J. Chromatog.* **615,** 1–22.
4. Bergman, T., Carlquist, M., and Jörnvall, H. (1986) Amino acid analysis by high performance liquid chromatography of phenylthiocarbamyl derivatives, in: *Advanced Methods in Protein Microsequence Analysis* (Wittmann-Liebold, B., Salnikov, J., and Erdmann, V. A., eds.), Springer, Berlin/Heidelberg, pp. 45–55.
5. Stark, M., Wang, Y., Danielsson, O., Jörnvall, H., and Johansson, J. (1998) Determination of proteins, phosphatidylethanolamine, and phosphatidylserine in organic solvent extracts of tissue material by analysis of phenylthiocarbamyl derivatives. *Anal. Biochem.* **265,** 97–102.
6. Bidlingmeyer, B. A., Cohen, S. A., and Tarvin, T. L. (1984) Rapid analysis of amino acids using pre-column derivatization. *J. Chromatog.* **336,** 93–104.
7. Koop, D. R., Morgan, E. T., Tarr, G. E., and Coon, M. J. (1982) Purification and characterization of a unique isozyme of cytochrome P-450 from liver microsomes of ethanol-treated rabbits. *J. Biol. Chem.* **257,** 8472–8480.

14

Analysis of *O*-Phosphoamino Acids in Biological Samples by Gas Chromatography with Flame Photometric Detection

Hiroyuki Kataoka, Norihisa Sakiyama, Yukizo Ueno, Kiyohiko Nakai, and Masami Makita

1. Introduction

The separation and determination of *O*-phosphoamino acids has been carried out by thin-layer chromatography *(1–8)*, thin-layer electrophoresis *(1,6,7,9–14)*, gel electrophoresis *(15)*, amino acid analyzer *(16–18)*, high-performance liquid chromatography (HPLC) *(19–33)*, capillary zone electrophoresis (CZE) *(34,35)* and immunoassay *(36–38)*. However, most of these methods require ^{32}P-labeling for detecting and quantifying *O*-phosphoamino acids, and therefore cannot be used for analysing nonradiolabeled *O*-phosphoamino acids in biological samples. Furthermore, some of them show poor resolution or low sensitivity. HPLC methods based on the precolumn ultraviolet (UV) derivatization with phenyl isothiocyanate *(22–24)*, dabsyl chloride *(25)*, and N-α-(2,4-dinitro-5-fluorophenyl)-l-alaninamide *(26)*, the postcolumn fluorescence derivatization with *o*-phthalaldehyde *(27–31)*, the precolumn fluorescence derivatization with 9-fluorenylmethyl chloroformate *(32,33)*, and CZE methods based on the UV derivatization with phenyl isothiocyanate or dabsyl chloride *(34,35)* were highly sensitive, but some of these methods lack specificity. Furthermore, many of these methods require lengthy separation time, and clean-up of the sample to remove the excess reagent and coexisting substances. On the other hand, the immunoassays by antibodies raised against each *O*-phosphoamino acid have been employed for measuring endogeneously phosphorylated amino acids, but these methods involve some problems in the specificity and crossreactivity of the antibodies. Current methods for the deter-

From: *Methods in Molecular Biology, vol. 159: Amino Acid Analysis Protocols*
Edited by: C. Cooper, N. Packer, and K. Williams © Humana Press Inc., Totowa, NJ

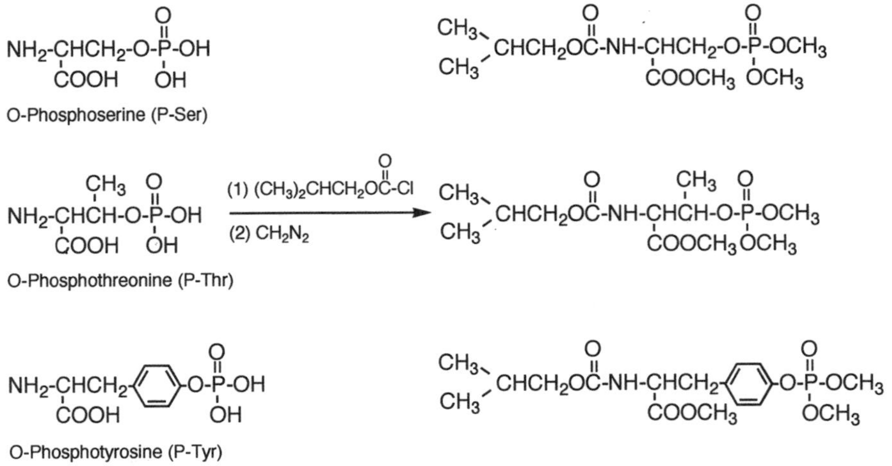

Fig. 1. Derivatization process of O-phosphoamino acids.

mination of O-phosphoamino acids have also been described in detail in **ref. 39**, and the review in **ref. 40**.

Recently, we have developed a selective and sensitive method for the determination of O-phosphoamino acids, such as O-phosphoserine (P-Ser), O-phosphothreonine (P-Thr), and O-phosphotyrosine (P-Tyr) by GC with flame photometric detection (FPD) using a capillary column, in which these compounds were analyzed as their N-isobutoxycarbonyl (isoBOC) methyl ester derivatives *(41–49)*. By using this method, we demonstrated that the contents of free O-phosphoamino acids in tissues *(42,43)* and urine *(48)*, and protein-bound O-phosphoamino acids in phosphorylated proteins *(44,45)*, tissues *(46,49)*, and urine *(47–49)* could be rapidly and simply analyzed. In this chapter, selective and sensitive methods for the determination of O-phosphoamino acids in biological samples by FPD-GC are described on the basis of the aforementioned results, and optimum conditions and typical problems encountered in the development and application of the methods are discussed.

The derivatization process is shown in **Fig. 1**. N-isoBOC methyl esters of O-phosphoamino acids can be easily, rapidly, and quantitatively prepared by the reaction with isobutyl chloroformate (isoBCF) by shaking in aqueous alkaline media, followed by esterification with diazomethane. The O-phosphoamino acids in peptides or proteins were released from the sample by acid and base hydrolyzes prior to derivatization. The main advantage of this method is that these amino acids can be easily converted in an aqueous medium without any further clean-up procedure, and the derivatives are stable to moisture. Furthermore, the derivatives can be quantitatively and reproducibly resolved as single peaks using a capillary column, and provide an excellent FPD response.

2. Materials

2.1. Equipment

1. Shimadzu 14A gas chromatograph equipped with a flame ionization detector and a flame photometric detector (P-filter).
2. For the analysis of free *O*-phosphoamino acids, fused silica capillary columns (15 m × 0.53-mm id, 1.0-μm film thickness) of crosslinked DB-1701 (14% cyanopropylphenyl-86% methylpolysiloxane, J & W, Folsom, CA) or DB-210 (50% trifluoropropyl-50% methylpolysiloxane, J & W).
3. For the analysis of protein-bound P-Ser/P-Thr and P-Tyr, DB-1707 (15 m × 0.53-mm id, 1.0-μm film thickness) and DB-5 (5% phenyl-95% methylpolysiloxane, J & W, Folsom, CA, USA: 15 m × 0.53-mm id, 1.0-μm film thickness).
4. Model LK-21 ultradisperser (Yamato Kagaku, Tokyo, Japan).
5. Model RD-41 Centrifugal evaporator (Yamato Kagaku).
6. Pico-Tag workstation (Waters Associates, Milford, MA).
7. Pasteur capillary pipet (Iwaki glass No. IK-PAS-5P).

2.2. Reagents

1. Standard *O*-phosphoamino acids, aminophosphonic acids and nonphosphorylated amino acids: *O*-phospho-L-serine (P-Ser), *O*-phospho-D,L-threonine (P-Thr), *O*-phospho-L-tyrosine (P-Tyr), *O*-phosphoethanolamine (PEA), and 2-aminoethylphosphonic acid (AEP) were purchased from Sigma (St. Louis, MO). All of the nonphosphorylated amino acids used were purchased from Ajinomoto (Tokyo, Japan).
2. Internal standards (IS): 2-amino-4-phosphonobutyric acid (APB), 2-amino-7-phosphonoheptanoic acid (APH) and 2-amino-8-phosphonooctanoic acid (APO) as internal standards (IS) were purchased from Sigma.
3. Protein samples: purified commercial proteins, bovine serum albumin (Fraction V, 96–99%), hen egg ovalbumin (Type VII), bovine milk α-casein, bovine milk β-casein, bovine milk κ-casein, egg yolk phosvitin, human hemoglobin, bovine heart cytochrome *c*, salmon sperm protamine, calf thymus histone (Type II), bovine brain myelin basic protein, yast alcohol dehydrogenase, bovine liver catalase, egg white lysozyme, and bovine pancreas chymotrypsin were purchased from Sigma.
4. Derivatizing reagents: isobutyl chloroformate (isoBCF) obtained from Tokyo Kasei Kogyo was used without further purification and stored at 4°C when not in use. *N*-methyl-*N*-nitroso-*p*-toluenesulphonamide and diethyleneglycol monomethyl ether for the generation of diazomethane *(50)* were obtained from Nacalai Tesuque.
5. Other materials: peroxide-free diethyl ether was purchased from Dojindo Laboratories (Kumamoto, Japan). Distilled water was used after fresh purification with a Model Milli-Q Jr. water purifier (Millipore, Bedford, MA). All other chemicals were analytical grade.

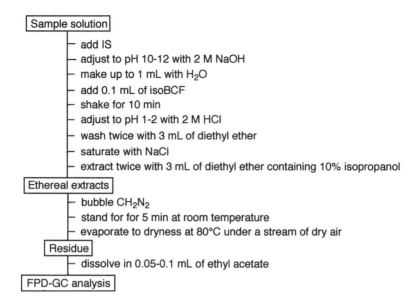

Fig. 2. Shematic flow diagram of the derivatization procedure.

2.3. Solutions

1. Standard *O*-phosphoamino acid solutions: each compound is dissolved in distilled water to make up a stock solution at a concentration of 0.1 mg/mL and then stored at 4°C. The working standard solutions are made up freshly, as required by dilution of the stock solution with distilled water. These solutions are stable at 4°C for at least 2 wk.
2. Internal standard solutions: APB, APH, and APO are dissolved in distilled water to make up stock solutions at a concentration of 0.1 mg/mL. These solutions are stable at 4°C for at least 2 wk.
3. Trichloroacetic acid (TCA) solution: TCA is dissolved in distilled water at concentrations of 5% and 20%.

3. Method
3.1. Derivatization

Derivatization procedure of *O*-phosphoamino acids is shown in **Fig. 2** (*see* **Note 1**).

1. For the analysis of free *O*-phosphoamino acids, pipet an aliquot of a sample containing 10–2000 ng of each amino acid and 0.1 mL of 2.5 µg/mL APB (IS) (if necessary) into a 10-mL Pyrex glass reaction tube with a PTFE-lined screw cap.
2. For the analysis of *O*-phosphoamino acids in protein hydrolysates, pipet an aliquot of a sample containing 10–500 ng of each amino acid and 0.05–0.1 mL of 1

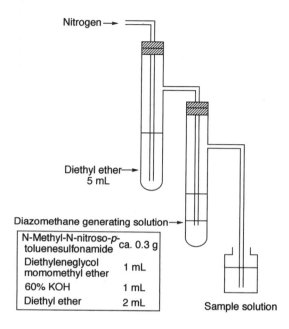

Fig. 3. Apparatus for microscale generation of diazomethane.

μg/mL APB (P-Ser/P-Thr analysis) and APH or APO (for P-Tyr analysis) (IS) (if necessary) into a reaction tube.

3. Adjust these mixtures to pH 10.0–12.0 with 2 *M* NaOH and make the total volume up to 1 mL with distilled water.

4. Immediately add 0.1 mL of isoBCF to the mixtures and shake at 300 cycles/min (up and down) for 10 min at room temperature (*see* **Note 2**).

5. Readjust the reaction mixture to pH 1.0–2.0 with 2 *M* HCl.

6. Extract twice with 3 mL of diethyl ether to remove the excess of reagent. Shake vigorously for 5–10 s by hand, then centrifugation at 2000*g* for 30 s (*see* **Note 3**). The ethereal extract can then be discarded.

7. Saturate the aqueous layer with NaCl and extract twice with 3 mL of diethyl ether containing 10% of isopropanol with vigorous shaking for 5–10 s by hand. After centrifugation at 2000*g* for 30 s, the organic layers should be transferred into another tube by means of a Pasteur capillary pipet (*see* **Note 4**).

8. The pooled ethereal extracts are methylated by bubbling diazomethane, generated according to the microscale procedure *(50)* (*see* **Note 5**). As shown in **Fig. 3**, a stream of nitrogen is saturated with diethyl ether in the first side-arm test tube and then passed through a diazomethane generating solution containing *N*-methyl-*N*-nitroso-*p*-toluenesulphonamide, diethylene-glycol monomethyl ether, and KOH. The generated diazomethane is carried into the sample tube via the nitrogen stream until a yellow tinge becomes visible. Leave to stand for more than 5

min at room temperature. The solvents are then removed by evaporation to dryness at 80°C under a stream of dry air. The residue on the walls of the tube are dissolved in 0.05–0.1 mL of ethyl acetate and then 1 μL of this solution is injected into the gas chromatograph (*see* **Note 6**).

3.2. Gas Chromatography

1. DB-1701 (14% cyanopropylphenyl-86% methylpolysiloxane, J & W, GC analysis was carried out with a Shimadzu 14A gas chromatograph equipped with a flame ionization detector and a flame photometric detector (P-filter).
2. For the analysis of free *O*-phosphoamino acids, fused-silica capillary columns (15 m × 0.53-mm id, 1.0-μm film thickness) of crosslinked DB-1701 or DB-210 should be used (*see* **Note 7**).
3. For a DB-1701 column, the operating conditions are as follows: column temperature program at 5°C/min from 180°C to 280°C; injection and detector temperature, 290°C.
4. For a DB-210 column, the operating conditions are as follows: column temperature program at 3°C/min from 200°C to 260°C; injection and detector temperature, 270°C.
5. For the analyzes of protein-bound P-Ser/P-Thr and P-Tyr, DB-1701 (15 m × 0.53-mm id, 1.0-μm film thickness) and DB-5 (15 m × 0.53-mm id, 1.0-μm film thickness) should be used (*see* **Notes 8** and **9**).
6. For the *O*-phosphoamino acid analysis of phosphorylated protein samples, a DB-1701 column is used with the following operating conditions: column temperatures, 210°C (for P-Ser/P-Thr) and 270°C (for P-Tyr); injection and detector temperatures, 260°C (P-Ser/P-Thr) and 290°C (for P-Tyr) (*see* **Note 10**).
7. For the P-Ser/P-Thr analyzes of protein-bound urine and tissue samples, a DB-1701 column is used with the following operating conditions: column temperature, isothermal at 220°C; injection and detector temperature, 260°C (*see* **Note 11**).
8. For the P-Tyr analyzes of protein-bound urine and tissue samples, a DB-5 column should be used with the following operating conditions: column temperature program at 3°C/min from 230°C to 280°C; injection and detector temperature, 290°C (*see* **Note 11**).
9. The nitrogen flow rate in each of these GC analyzes is 10 mL/min.
10. Measure the peak heights of each *O*-phosphoamino acid and the IS. Calculate the peak height ratios against the IS to construct calibration curves (*see* **Notes 12–14**).

3.3. Tissue Sample Preparation

3.3.1. Preparation of Tissue Samples
for the Free O-Phosphoamino Acid Analysis

Preparation of tissue samples for the free *O*-phosphoamino acid analysis *(43)*: the following species were used in this experiment: cuttlefish (*Ommastrephes sloani pacificus*), mackerel (*Scomber Japonicus*), chicken, mouse, and pig (*see* **Note 15**).

1. After dissection of each animal, remove the organs and chill on ice. Chop the organs up into pieces and then store frozen at –20°C until the assay.
2. Extract the *O*-phosphoamino acids from a tissue sample within 48 h by the following procedure, the assay being repeated three times.
3. Homogenize each tissue (0.2–1 g) with 5 vol of 5% of TCA using a Model LK-21 ultradisperser (Yamato Kagaku, Tokyo, Japan).
4. Centrifuge at 2000g for 10 min, then reextract the precipitate with 5 vol of 5% of TCA (*see* **Note 16**).
5. The combined supernatants are clarified by passage through a glass filter, if necessary, washed three times with two volumes of diethyl ether to remove TCA, and then warmed at 60°C to remove the ether.
6. After cooling, the solution is adjusted to pH 6.0–7.0 by the addition of a few drops of 1 *M* NaOH and then run through a Dowex 1 × 8 (4 × 0.7-cm id, AcO⁻ form, 100–200 mesh) column (*see* **Note 17**).
7. Wash the column with 15 mL of water and then elute with 0.1 *M* HCl. The initial 9 mL of the eluate is discarded and the following 5 mL of the eluate (pH 1.0–2.0) is collected as the *O*-phosphoamino acid fraction. To this fraction, add 0.1 mL of 2.5 μg/mL APB (IS) solution, and the resultant mixture is used for the derivatization.

3.3.2. Preparation of Tissue Samples for the Protein-Bound O-Phosphoamino Acid Analysis

Preparation of tissue samples for the protein-bound *O*-phosphoamino acid analysis *(46)*: seven male ICR mice (6 wk old) were used in the experiments (*see* **Notes 18** and **19**).

1. Immediately after dissection, remove each organ and store at –20°C until used. Chop up each pooled tissue. An aliquot (approx 0.5 g) should be homogenized with 5 vol of 5% TCA using an ultradisperser.
2. Centrifuge at 2000g for 10 min and reextract the precipitate with 5 vol of 5% TCA.
3. Wash the resulting pellet twice with 5 mL of diethyl ether to remove the TCA, then warm to dryness at 60°C (*see* **Note 20**).
4. Weigh the residue, then ground it to a powder, this being called the protein-bound fraction.
5. Hydrolyze an aliquot (5–10 mg) of the powdered protein-bound fraction with acid and base, as described in **Subheading 3.4.**
6. Adjust the pH of the acid and base hydrolyzates to pH 10.0–12.0 with 2 *M* NaOH or 10 *M* HCl, as described in **Subheading 3.4.**
7. The samples are then ready for derivatization (*see* **Notes 21** and **22**).

3.3.3. Preparation of Urine Samples for the Free O-Phosphoamino Acid Analysis

Preparation of urine samples for the free *O*-phosphoamino acid analysis *(48)*.

1. Collect 24-h urine samples from healthy volunteers under toluene and keep frozen, if not analyzed immediately.

2. To 3 mL of urine samples, add 1 mL of 20% TCA and mix.
3. Centrifugation at 2000g for 10 min.
4. Wash the supernatant three times with 3 mL of diethyl ether to remove the TCA, and warm at 60°C to remove the ether.
5. After cooling, the solution should be adjusted to pH 6.0–7.0 by the addition of a few drops of 1 M NaOH and then run through a Dowex 1 × 8 (4.X 0.7-cm id, AcO⁻ form, 200–400 mesh) column. Wash the column with 15 mL of water and then elute with 0.1 M HCl. Discard the initial 7.5 mL of the eluate. Collect the following 5 mL of the eluate (pH 1.0–2.0) as the O-phosphoamino acid fraction.
6. To this fraction, add 0.05 mL of 1 μg/mL APB and APO (IS) solutions.
7. The resultant mixtures are used for the derivatization.

3.3.4. Preparation of Urine Samples for the Total O-Phosphoamino Acid Analysis

Preparation of urine samples for the total (free plus protein-bound) O-phosphoamino acid analysis *(48)* (*see* **Notes 23** and **24**):

1. Collect, under toluene, 24-h urine samples from healthy volunteers and keep frozen, if not analyzed immediately.
2. Perform acid and base hydrolyzes of urine samples, as described in **Subheading 3.4.**
3. For the P-Ser/P-Thr analysis, add 0.1 mL of 1 μg/mL APB (IS) to 0.1 mL of the urine sample and evaporated to dryness at 60°C in a centrifugal evaporator. Hydrolyzed the residue with 0.2 mL of 6 M HCl containing 1% phenol in the vapor phase for 2 h at 110°C under vacuum in a Pico-Tag workstation.
4. For the P-Tyr analysis, add 0.05 mL of 1 μg/mL APO (IS) and 1.22 mL of 10.5 M KOH to 3 mL of the urine sample and hydrolyze for 1 h at 130°C in a Pico-Tag workstation.
5. Adjust the resulting acid and base hydrolysates to pH 10.0–12.0 with 2 M NaOH or 10 M HCl, ready for derivatization.

3.4. Protein Sample Preparation

3.4.1 Acid Hydrolysis

For P-Ser and P-Thr analysis the proteins are hydrolyzed in an acid (*see* **Notes 25** and **26**).

1. Place an aliquot of the protein sample (0.004–2 mg) in a 5 × 50-mm glass test tube, add 50 μL of 20 μM APB (IS). Dry the mixture in a centrifugal evaporator.
2. Hydrolyze the residue with 0.2 mL of 6 M HCl containing 1% phenol in the vapor phase for 2 h at 110°C under vacuum in a Pico-Tag workstation *(51)*.
3. Extract the resulting hydrolysate twice with 0.5 mL of distilled water, the extracts being transferred to another reaction tube (10-mL Pyrex glass tube with a PTFE-lined screw cap), and then used for derivatization.

3.4.2. Base Hydrolysis

For P-Tyr analysis, the proteins are hydrolyzed in a base (*see* **Note 27**).

1. Place an aliquot of the protein sample (0.2–4 mg) in a 9 × 75-mm polypropylene tube, add 50 µL of 4 µ*M* APH (IS).
2. Add 0.2 mL of 10 *M* KOH, and make the total volume up to 0.4 mL with distilled water, if necessary.
3. Hydrolyze the mixture by heating for 1 h at 130°C under vacuum in a Pico-Tag workstation.
4. Add 0.2 mL of 8 *M* HCl to the resulting hydrolysate and transfer the mixture to another reaction tube. This is now ready for derivatization.

4. Notes

1. IsoBCF and diazomethane are used as a derivatizing reagents for the amino group and the carboxyl and phosphoryl groups, respectively (**Fig. 1**).
2. The *N*-isobutoxycarbonylation of *O*-phosphoamino acids with isoBCF proceeds rapidly and quantitatively in aqueous alkaline media (*41*). This reaction is completed with >50 µL of isoBCF within 5 min by shaking at room temperature.
3. Although diethyl ether has been used as an extraction solvent for *N*-isoBOC aminocarboxylic acids (*52,53*), *N*-isoBOC phosphoamino acids cannot be extracted with this solvent. Therefore, the reaction mixture is washed with diethyl ether under acid condition in order to remove not only the excess reagent, but also the amines, phenols, and aminocarboxylic acids, which often coexist with *O*-phosphoamino acids in biological samples and derivatize to corresponding *N*- and *O*-isoBOC derivatives under same conditions as described in **Note 2**.
4. In order to extract the remaining *N*-isoBOC *O*-phosphoamino acids in aqueous layer, the addition of another solvent to diethyl ether was tested. Among the various solvents tested, isopropanol proved to be the most satisfactory solvent for this purpose, and its optimum concentration was found to be in the range 5–15%. Thus, the *N*-isoBOC phosphoamino acids in aqueous layer could be quantitatively and selectively extracted into diethyl ether containing 10% isopropanol in acid condition with NaCl saturation. In this procedure, the organic layers should be collected, taking care to avoid aqueous droplets. It was not necessary to complete draw the organic layer in each extraction.
5. The methylation of the ethereal extracts is successfully carried out by bubbling diazomethane. This reaction should be performed in a well-ventilated hood because diazomethane is explosive and toxic.
6. The derivative preparation can be performed within 30 min, and several samples can be treated simultaneously. The N-isoBOC methyl ester derivatives of *O*-phosphoamino acids are very stable to moisture; therefore, no precaution to exclude moisture is necessary in their handling and storage. No decomposition was observed during GC analysis and even after standing in ethyl acetate for 3 wk at room temperature.

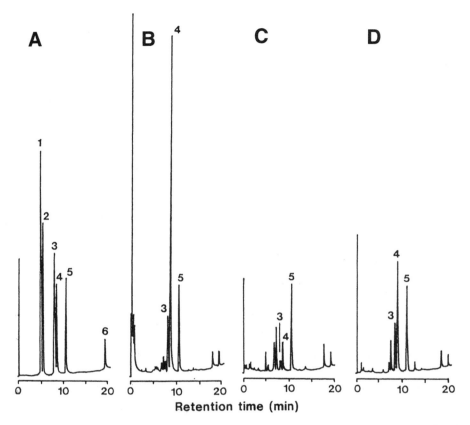

Fig. 4. Chromatograms obtained from standard and animal tissue samples for the free *O*-phosphoamino acid analysis by FPD-GC. **(A)** Standard (containing 0.25 µg of each amino acid); **(B)** mackerel spleen (340 mg); **(C)** mouse brain (744 mg); **(D)** pig stomach (596 mg). GC conditions: column, DB-1701 (15 m × 0.53-mm id, 1.0-µm film thickness); column temperature, programmed at 5°C/min from 180°C to 280°C; injection and detector temperatures, 290°C; nitrogen carrier gas flow rate, 10 mL/min. Peaks: 1 = 2-aminoethylphosphonic acid (AEP); 2 = *O*-phosphoethanolamine (PEA); 3 = *O*-phosphothreonine (P-Thr); 4 = *O*-phosphoserine (P-Ser); 5 = 2-amino-4-phosphonobutyric acid (APB) (IS); 6 = *O*-phosphotyrosine (P-Tyr).

7. The GC analysis was performed with FPD (526-nm interference filter), which is highly selective for phosphorus containing compounds, by using a megabore capillary column and direct (splitless) sample injection system. In this system, the sufficiently inactivated glass insert should be used to avoid tailing peaks of derivatives caused by the adsorption on injection port.

8. In preliminary test for several megabore capillary columns, DB-17, DB-1701, DB-210, and DB-5 gave good separation for *O*-phosphoamino acids.

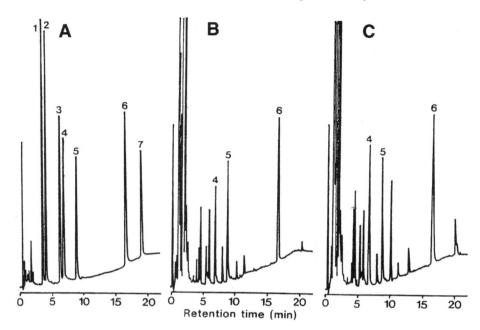

Fig. 5. Chromatograms obtained from standard and urine samples for the free *O*-phosphoamino acid analysis by FPD-GC. (A) Standard (containing 50 ng of each amino acid); (B) and (C) urine (3 mL) samples. GC conditions: column, DB-210 (15 m × 0.53-mm id, 1.0-μm film thickness); column temperature, programmed at 3°C/min from 200°C to 260°C; injection and detector temperatures, 270°C; nitrogen carrier gas flow rate, 10 mL/min. Peaks: 1 = 2-aminoethylphosphonic acid (AEP); 2 = *O*-phosphoethanolamine (PEA); 3 = *O*-phosphothreonine (P-Thr); 4 = *O*-phosphoserine (P-Ser); 5 = 2-amino-4-phosphonobutyric acid (APB) (IS); 6 = 2-amino-8-phosphonooctanoic acid (APO) (IS); 7 = *O*-phosphotyrosine (P-Tyr) (from **ref. 48** with permission).

9. For the free *O*-phosphoamino acid analysis, each amino acid gives a single and symmetrical peak from FPD-GC with a DB-1701, and are completely separated within 16 min from other biological phosphorus amino compounds, such as PEA and AEP (**Fig. 4A**) *(43)*. Furthermore, the free *O*-phosphoamino acids can be also separated from these compounds using a DB-210 (**Fig. 5A**) *(48)*. Other phosphate compounds, such as sugar phosphate and nucleic acid phosphate, and nonphosphorus amino acids were not detected by this method.

10. For the P-Ser/P-Thr and P-Tyr analyzes of phosphorylated protein samples, DB-1701 was used under different GC conditions, because these samples were separately treated by acid and base hydrolyzes. As shown in **Figs. 6A** and **7A**, P-Ser/P-Thr and P-Tyr samples were separated within 12 and 8 min, respectively *(44)*.

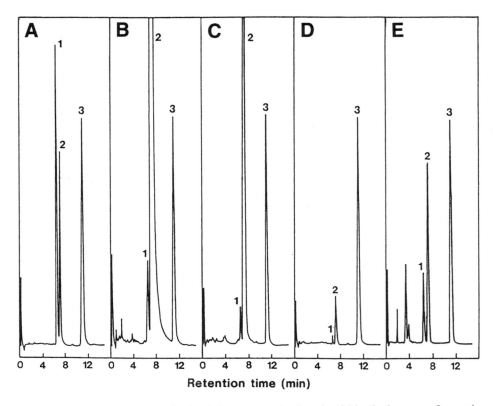

Fig. 6. Chromatograms obtained from a standard and acid hydrolysates of protein samples by FPD-GC. **(A)** Standard (containing 1 nmol of each compound); **(B)** α-casein (0.5 mg); **(C)** phosvitin (0.01 mg); **(D)** histone (0.8 mg); **(E)** myelin (0.25 mg). GC conditions: column, DB-1701 (15 m × 0.53-mm id, 1.0-μm film thickness); column temperature, 210°C; injection and detector temperatures, 260°C; nitrogen carrier gas flow rate, 10 mL/min. Peaks: 1 = *O*-phosphothreonine (P-Thr); 2 = *O*-phosphoserine (P-Ser); 3 = 2-amino-4-phosphonobutyric acid (APB) (IS) (from **ref.** *44* with permission).

11. DB-1701 and DB-5 proved to be the most satisfactory columns for the P-Ser/P-Thr and P-Tyr analyzes of urine hydrolysates, respectively. By using these columns, each *O*-phosphoamino acid elutes separately as single and symmetrical peaks within 10 min (**Figs. 8A** and **9A**) *(48)*.

12. The calibration curves for the free *O*-phosphoamino acids by FPD-GC were conducted using APB, which showed a similar behavior to other *O*-phosphoamino acids during the derivatization and was well separated from these amino acids on a chromatogram as the IS. A linear relationship was obtained at the range 0.01–2 mg with the regression lines for P-Ser, P-Thr and P-Tyr being $y = 3.938x - 0.040$ ($r = 0.9984$, $n = 18$), $y = 5.183x - 0.028$ ($r = 0.9983$, $n = 18$), and $y = 1.761x -$

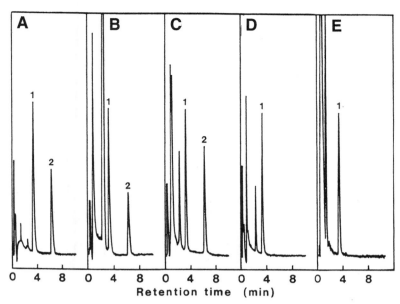

Fig. 7. Chromatograms obtained from a standard and base hydrolysates of protein samples by FPD-GC. (A) Standard (containing 0.2 nmol of each compound); (B) α-casein (2.0 mg); (C) phosvitin (1.3 mg); (D) histone (2.7 mg); (E) myelin (1.0 mg). GC conditions: column, DB-1701 (15 m × 0.53-mm id, 1.0-μm film thickness); column temperature, 270°C; injection and detector temperatures, 290°C; nitrogen carrier gas flow rate, 10 mL/min. Peaks: 1 = 2-amino-7-phosphonoheptanoic acid (APH) (IS); 2 = O-phosphotyrosine (P-Tyr) (from **ref. 42** with permission).

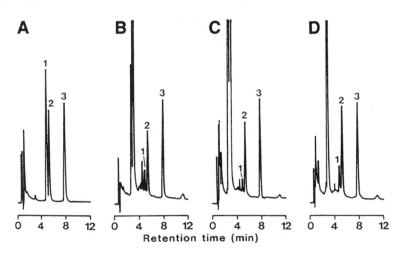

Fig. 8. Chromatograms obtained from a standard and acid hydrolysates of urine samples by FPD-GC. ; Standard (containing 100 ng of each compound); (B)–(D) urine (0.1 mL) hydrolysates. GC conditions: column, DB-1701 (15 m × 0.53-mm id, 1.0-μm film thickness); column temperature, 220°C; injection and detector temperatures, 260°C; nitrogen carrier gas flow rate, 10 mL/min. Peaks: 1 = O-phosphothreonine (P-Thr); 2 = O-phosphoserine (P-Ser); 3 = 2-amino-4-phosphonobutyric acid (APB) (IS) (from **ref. 48** with permission).

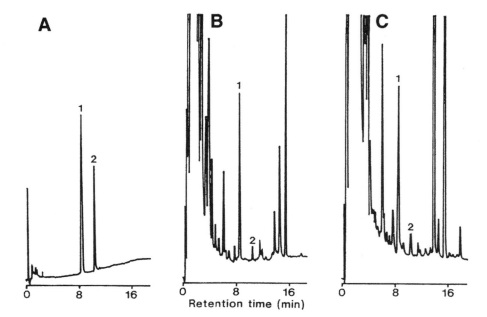

Fig. 9. Chromatograms obtained from a standard and base hydrolysates of urine samples by FPD-GC. **(A)** Standard (containing 50 ng of each compound); **(B)** and **(C)** urine (3 mL) hydrolysates. GC conditions: column, DB-5 (15 m × 0.53-mm id, 1.0-μm film thickness); column temperature, programmed at 3°C/min from 230°C to 280°C; injection and detector temperatures, 290°C; nitrogen carrier gas flow rate, 10 mL/min. Peaks: 1 = 2-amino-8-phosphonooctanoic acid (APO) (IS); 2 = *O*-phosphotyrosine (P-Tyr) (from **ref. *48*** with permission).

0.054 ($r = 0.9972$, $n = 18$), respectively, where y is the peak height ratio and x is the amount (μg) of each compound.

13. The derivatives provided excellent FPD responses and the minimum detectable amounts of P-Ser, P-Thr, and P-Tyr at a signal three times as high as the noise under our instrumental conditions were approx 50, 40, and 200 pg as injection amounts, respectively. The FPD-GC system described here was over 200 times more sensitive than the FID-GC system.

14. The calibration curves for the analysis of *O*-phosphoamino acids in protein hydrolysates were conducted using APB (for P-Ser/P-Thr analysis) and APH or APO (for P-Tyr analysis) as the IS *(44,48)*. In each case, a linear relationship was obtained using different GC columns in the range 10–500 ng of each *O*-phosphoamino acid and the correlation coefficients were above 0.994 *(48)*. The detection limits for P-Ser, P-Thr, and P-Tyr in these GC conditions were approx 40, 30, and 80 pg as injection amounts, respectively.

15. The method developed for free *O*-phosphoamino acids was successfully applied to animal tissue *(43)* and urine *(48)* samples. In preliminary experiments on mice,

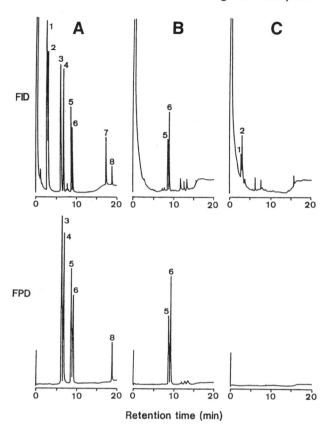

Fig. 10. Chromatograms obtained from (**A**) a standard solution, and (**B**) nonhydrolyzed and (**C**) hydrolyzed samples of porcine liver. The converted samples were analyzed by both FID-GC (top) and FPD-GC (bottom). GC conditions: column, DB-1701 (15 m × 0.53-mm id, 1.0-μm film thickness); column temperature, isothermal at 180°C for 6 min, programmed at 20°C/min to 220°C, isothermal at 220°C for 6 min, programmed at 20°C/min to 270°C, and isothermal at 270°C for 5 min; injection and detector temperatures, 280°C; nitrogen carrier gas flow rate, 10 mL/min. Peaks: 1 = threonine (Thr); 2 = serine (Ser); 3 = 2-aminoethylphosphonic acid (AEP); 4 = *O*-phosphoethanolamine (PEA); 5 = *O*-phosphothreonine (P-Thr); 6 = *O*-phosphoserine (P-Ser); 7 = tyrosine (Tyr); 8 = *O*-phosphotyrosine (P-Tyr) (from **ref. 42** with permission).

no postmortem change of *O*-phosphoamino acids was observed after 24 h. *O*-phosphoamino acids in various organs were stable for at least 2 d under freezing at –20°C.
16. TCA is routinely used to precipitate proteins with no loss of phosphoryl groups *(54)*. Free *O*-phosphoamino acids in tissue and urine samples were quantitatively extracted into the TCA soluble fraction.

Table 1
Recoveries of *O*-Phosphoamino Acids Added
to Tissue and Urine Samples

Sample	*O*-Phospho-amino acid	Added[a]	Amount found[b] Nonaddition	Amount found[b] Addition	Recovery (%)
Pig tissue					
Brain	P-Ser	500	330 ± 20	850 ± 14	104.0
	P-Thr	100	57 ± 2	152 ± 3	95.0
	P-Tyr	100	ND[c]	103 ± 4	103.0
Heart	P-Ser	500	161 ± 11	671 ± 50	102.0
	P-Thr	100	18 ± 1	117 ± 3	99.0
	P-Tyr	100	ND	97 ± 2	97.0
Liver	P-Ser	500	377 ± 13	878 ± 64	100.2
	P-Thr	100	115 ± 2	214 ± 4	99.0
	P-Tyr	100	ND	102 ± 3	102.0
Kidney	P-Ser	500	567 ± 27	1085 ± 57	103.6
	P-Thr	100	46 ± 3	146 ± 8	100.0
	P-Tyr	100	ND	103 ± 5	103.0
Urine					
A	P-Ser	20	26.8 ± 0.7	46.4 ± 2.1	98.0
	P-Thr	20	ND	19.6 ± 0.6	98.0
	P-Tyr	20	ND	19.0 ± 1.1	95.0
B	P-Ser	20	12.3 ± 0.5	31.8 ± 0.9	97.5
	P-Thr	20	ND	18.4 ± 0.7	92.0
	P-Tyr	20	ND	18.7 ± 1.5	93.5
Urine hydrolysate[d]					
A	P-Ser	500	1017 ± 42	1504 ± 54	97.4
	P-Thr	500	133 ± 9	593 ± 35	92.0
	P-Tyr	50	2.4 ± 0.1	47.7 ± 0.7	90.6
B	P-Ser	500	1636 ± 60	2111 ± 118	95.0
	P-Thr	500	238 ± 7	694 ± 25	91.2
	P-Tyr	50	6.2 ± 0.4	53.5 ± 1.7	94.6

[a]Tissue sample, ng/g; urine sample, ng/mL.
[b]Tissue sample, ng/g; urine sample, ng/mL; mean $\pm$ SD (n=3).
[c]Not detectable.
[d]P-Ser and P-Thr were added to acid hydrolysate and P-Tyr was added to base hydrolysate.

17. The TCA extract was applied on a Dowex 1 column in order to remove any coexisting substances. *O*-phosphoamino acids were adsorbed on the column and were eluted in 0.1 *M* HCl fraction. This column chromatographic procedure could be completed within 50 min, and the recoveries of P-Ser, P-Thr, and P-Tyr were 97–103%. **Figs. 4B–D** *(43)* and **5B,C** *(48)* show the chromatograms obtained from

Table 2
O-Phosphoamino Acid Contents in Several Commercial Proteins

O-Phosphoamino acid (nmol/mg of protein)[a]

Protein	P-Ser[b]	P-Thr[b]	P-Tyr[c]
Albumin	ND[d]	ND	ND
Ovalbumin	7.74 ± 0.51	ND	ND
α-Casein	137.30 ± 14.20	0.69 ± 0.04	0.073 ± 0.004
β-Casein	104.28 ± 4.87	0.16 ± 0.01	0.013 ± 0.001
κ-Casein	67.26 ± 0.89	0.87 ± 0.01	0.071 ± 0.002
Phosvitin	848.14 ± 35.81	11.06 ± 0.75	0.182 ± 0.010
Hemoglobin	0.03 ± 0.002	ND	ND
Cytochrome c	0.08 ± 0.002	ND	ND
Protamine	1.46 ± 0.07	ND	ND
Histone	0.30 ± 0.01	0.04 ± 0.002	ND
Myelin basic protein	3.93 ± 0.28	1.00 ± 0.06	ND
Alcohol dehydrogenase	0.07 ± 0.004	ND	ND
Catalase	0.17 ± 0.01	0.04 ± 0.001	ND
Lysozyme	0.05 ± 0.003	ND	ND
Chymotrypsin	0.13 ± 0.01	ND	ND

[a]Mean ± SD (n = 3).
[b]Analyzed after partial acid hydrolysis.
[c]Analyzed after partial base hydrolysis.
[d]Not detectable.
(From **ref. 44** with permission.)

animal tissue and urine samples, respectively. O-phosphoamino acids could be detected without any interference from coexisting substances by FPD-GC.

18. P-Ser and P-Thr were found in various animal tissues, but only P-Ser was found in urine samples. The occurrence of free P-Ser and P-Thr in these samples was confirmed by GC, GC-mass spectrometry (GC-MS), and thin-layer chromatography of nonhydrolyzed and hydrolyzed samples *(42)*. As shown in **Fig. 10**, peaks corresponding to P-Ser and P-Thr were observed in the chromatograms obtained by FID-GC (top), and FPD-GC (bottom) analyzes of the porcine liver sample. But these peaks disappeared after acid hydrolysis of sample with 6 M HCl at 110°C for 48 h, and the peaks corresponding to Ser and Thr newly appeared instead (**Fig. 10**). As shown in **Table 1**, the recoveries of P-Ser, P-Thr, and P-Tyr added to tissue and urine samples were 92–104%, the relative standard deviations being 1.6–8.0% (n = 3).

19. The method developed for protein-bound O-phosphoamino acids was successfully applied to purified proteins *(44)*, phosphorylated proteins by protein kinases *(45)*, animal tissue *(46)* and urine *(48)* samples. For the presence of O-phosphoamino acid residues in protein samples, several commercial proteins were analyzed by FPD-GC. As shown in **Figs. 6B–D** and **7B–D**, O-phosphoamino

acids in protein hydrolysates were selectively detected without any interference from other amino acids *(44)*. This method is reproducible and directly applicable for the analysis of the extent of both in vivo and in vitro phosphorylation without radiolabeling.

20. The proteins containing *O*-phosphoamino acids were recovered by better than 99% in the pellet with the TCA treatment without loss of phosphoryl groups *(54)*. Phospholipids, such as phosphatidylcholine and phosphatidylserine, were removed from the protein-bound fraction by washing the pellet with ether.

21. The results of *O*-phosphoamino acid analyzes of the protein samples after partial acid and base hydrolyzes are shown in **Table 2**. P-Ser was detected in all the samples investigated in this study, except for BSA, and P-Thr and P-Tyr were also detected in casein and phosvitin. This method was successfully applied to *O*-phosphoamino acid analysis of the proteins phosphorylated by protein kinases *(45)*.

22. Although the recovery of *O*-phosphoamino acids from the protein-bound fraction under the best conditions for acid hydrolysis was low (approx 35–40%) *(43)*, the method could reproducibly measure endogeneously phosphorylated amino acids in the tissue and urine samples without radiolabeling.

23. As shown in **Table 1**, the recoveries of *O*-phosphoamino acids added to urine hydrolysates were 91–97%, the relative standard deviations being 1.5–6.8% ($n = 3$).

24. **Figs. 8B–D** and **9B,C** show the chromatograms obtained from acid and base hydrolysates of urine samples by FPD-GC. The P-Ser, P-Thr, and P-Tyr peaks obtained from urine hydrolysates were confirmed by GC-MS analysis *(47)*. By using this method, we demonstrated that P-Tyr levels in mouse urine and liver increase during liver regeneration after partial hepatectomy *(49)*.

25. The O-phosphate linkages of Ser and Thr residues are defined by their stability to acids and lability to bases. Thus, acid hydrolysis has been routinely used to study the identity and amounts of P-Ser and P-Thr in peptides and proteins. For the acid hydrolysis of phosvitin (containing P-Ser, P-Thr, and P-Tyr residues) with the HCl vapor phase using a Pico-Tag workstation, the maximal recoveries of P-Ser and P-Thr were obtained at 110°C for 2–3 h (**Fig. 11**), these conditions being in agreement with those reported previously *(20)*. Under these conditions, P-Ser was detected as over 70 times more than P-Thr in phosvitin, whereas P-Tyr was not detected at all.

26. Under the best conditions for acid hydrolysis, the recovery of P-Ser from phosvitin was calculated to be $35.7 \pm 1.5\%$ ($n = 3$) by measuring the total phosphate content. The recovery of P-Ser from α-casein was $39.5 \pm 4.1\%$ ($n = 3$) under the same conditions. These low recoveries seem to have been caused by both incomplete peptide hydrolysis and destruction of the liberated P-Ser to Ser. The typical recovery for proteins has been reported to be in the range 20–30% *(19)*. Our results presented here are generally somewhat higher, the recoveries may show some dependence on the neighboring amino acid residues *(55)*.

27. The phosphate linkage of the Tyr residue is far more resistant to base than to acid hydrolysis. P-Tyr tolerated the base hydrolysis fairly well in 5 M KOH at 155°C

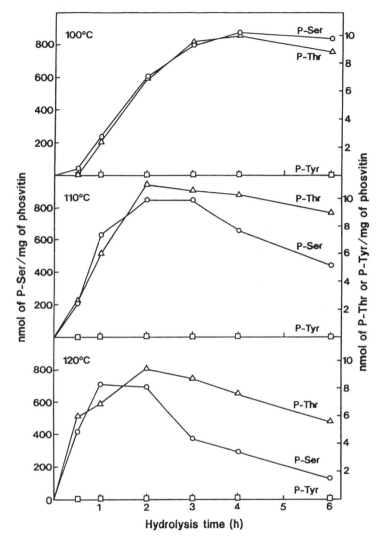

Fig. 11. Release of *O*-phosphoamino acids from phosvitin by acid hydrolysis.

for 30–35 min *(17,54)*. In our experiment, the maximal recovery of P-Tyr from phosvitin was obtained in 5 *M* KOH at 130°C for 0.5–4 h (**Fig. 12**).

References

1. Manai, M. and Cozzone, A. J. (1982) Two-dimential separation of phosphoamino acids from nucleoside monophosphates. *Anal. Biochem.* **124,** 12–18.
2. Chang, W.-C., Lee, M. L., Chou, C. K., and Lee, S. C. (1983) Polyamide thin-layer chromatography of phosphorylated tyrosine, threonine, and serine. *Anal. Biochem.* **132,** 342–344.

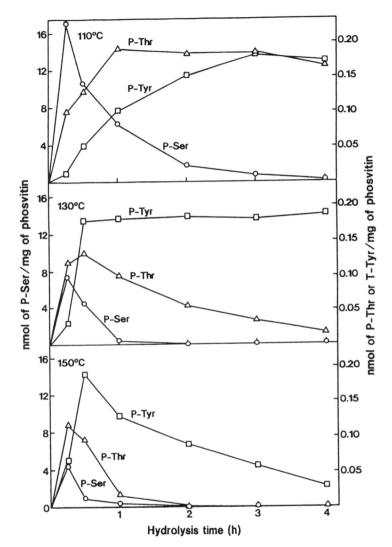

Fig. 12. Release of *O*-phosphoamino acids from phosvitin by base hydrolysis.

3. Fazekas, S., Ovary, I., and Szekessy-Hermann, V. (1989) Presence of phosphotyrosine in alkaline hydrolysate of pig skeletal muscle myosin. *Acta. Phys. Hung.* **74,** 161–168.
4. Neufeld, E., Goren, H. J., and Boland, D. (1989) Thin-layer chromatography can resolve phosphotyrosine, phosphoserine, and phosphothreonine in a protein hydrolyzate. *Anal. Biochem.* **177,** 138–143.
5. Munoz, G. and Marshall, S. H. (1990) An alternative method for a fast separation of phosphotyrosine. *Anal. Biochem.* **190,** 233–237.

6. Duclos, B., Marcandier, S., and Cozzone, A. J. (1991) Chemical properties and separation of phosphoamino acids by thin-layer chromatography and/or electrophoresis. *Methods Enzymol.* **201,** 10–21.
7. de Witte, P. A., Cuveele, J. F., Merlevede, W. J., and Vandenheede, J. R. (1995) Analysis of phosphorylhydroxyamino acids present in hydrolyzed cell extracts using dabsyl derivatization. *Anal. Biochem.* **226,** 1–9.
8. de Witte, P. A., Cuveele, J. F., Merlevede, W. J., and Agostinis, P. M. (1996) Analysis of the phosphoamino acid content of phosphoproteins. *J. Pharm. Biomed. Anal.* **14,** 1063–1067.
9. Kamps, M. P. and Sefton, B. M. (1989) Acid and base hydrolysis of phosphoproteins bound to immobilon facilitates analysis of phosphoamino acids in gel-fractionated proteins. *Anal. Biochem.* **176,** 22–27.
10. Hildebrandt, E. and Fried, V. A. (1989) Phosphoamino acid analysis of protein immobized on polyvinylidene difluoride membrane. *Anal. Biochem.* **177,** 407–412.
11. Lippmann, C., Lindschan, C., and Erdmann, V. A. (1992) Thin-layer electrophoresis with PhastSystem facilitates analysis of phosphoamino acids from proteins bound to Immobilon. *Electrophoresis* **13,** 666–668.
12. Durocher, Y. and Chevalier, S. (1994) Detection of phosphotyrosine in partial acid hydrolysis in gels. *J. Biochem. Biophys. Methods* **28,** 101-113.
13. Lombardini, J. B. and Props, C. (1995) Effects of cooling temperature on the separation of phosphoamino acids. *Anal. Biochem.* **227,** 399–400.
14. Mahoney, C. W., Nakanishi, N., and Ohashi, M. (1996) Phosphoamino acid analysis by semidry electrophoresis on cellulose thin-layer plates using the pharmacia/ LKB multiphor or Atto flatbed apparatus. *Anal. Biochem.* **238,** 96–98.
15. Yan, J. X., Packer, N. H., Tonella, L., Ou, K., Wilkins, M. R., Sanchez, J.-C., et al. (1997) High sample throughput phosphoamino acid analysis of proteins separated by one-and two-dimentional gel electrophoresis. *J. Chromatog. A* **764,** 201–210.
16. Kinnier, W. J. and Wilson, J. E. (1977) Complete analysis of protein hydrolysates containing phosphoserine and phosphothreonine using the amino acid analyzer. *J. Chromatog.* **135,** 508–510.
17. Martensen, T. M. (1982) Phosphotyrosine in proteins: Stability and quantification. *J. Biol. Chem.* **257,** 9648–9652.
18. Capony, J. P. and Demaille, J. G. (1983) A rapid microdetermination of phosphoserine, phosphothreonine, and phosphotyrosine in proteins by automatic cation exchange on a conventional amino acid analyzer. *Anal. Biochem.* **128,** 206–212.
19. Morrice, N. and Aitken, A. (1985) A simple amd rapid method of quantitative analysis of phosphoamino acids by high-performance liquid chromatography. *Anal. Biochem.* **148,** 207–212.
20. Robert, J. C., Soumarmon, A., and Lewin, M. J. M. (1985) Determination of O-phosphothreonine, O-phosphoserine, O-phosphotyrosine and phosphate by high-performance liquid chromatography. *J. Chromatog.* **338,** 315-324.
21. McCroskey, M. C., Colca, J. R., and Pearson, J. D. (1988) Determination of [^{32}P]phosphoamino acids in protein hydrolysates by isocratic anion-exchange high-performance liquid chromatography. *J. Chromatog.* **442,** 307–315.

22. Meyer, H. E., Swiderek, K., Hoffmann-Posorske, E., Korte, H., and Heilmeyer, L. M. G., Jr. (1987) Quantitative determination of phosphoserine by high-performance liquid chromatography as the phenylthiocarbamyl-S-ethylcycteine: application to picomolar amounts of peptides and proteins. *J. Chromatog.* **397,** 113–121.

23. Murthy, L. R. and Iqbal, K. (1991) Measurement of picomoles of phosphoamino acids by high-performance liquid chromatography. *Anal. Biochem.* **193,** 299–305.

24. Aebersold, R., Watts, J. D., Morrison, H. D., and Bures, E. J. (1991) Determination of the site of tyrosine phosphorylation at the low picomole level by automated solid-phase sequence analysis. *Anal. Biochem.* **199,** 51–60.

25. Malencik, D. A., Zhao, Z., and Anderson, S. R. (1990) Determination of dityrosine, phosphotyrosine, phosphothreonine, and phosphoserine by high-performance liquid chromatography. *Anal. Biochem.* **184,** 353–359.

26. Goodnough, D. B., Lutz, M. P., and Wood, P. L. (1995) Separation and quantification of D-and L-phosphoserine in rat brain using Na-(2, 4-dinitro-5-fluorophenyl)-L-alaninamide (Marfey's reagent) by high-performance liquid chromatography with ultraviolet detection. *J. Chromatog.* B, **672,** 290–294.

27. Steiner, A. W., Helandes, E. R., Fujitaki, J. M., Snith, L. S., and Smith, R. A. (1980) High-performance liquid chromatography of acid-stable and acid-labile phosphoamino acids. *J. Chromatog.* **202,** 263–269.

28. Yang, J. C., Fujitaki, J. M., and Smith, R. A. (1982) Separation of phosphohydroxyamino acids by high-performance liquid chromatography. *Anal. Biochem.* **122,** 360–363.

29. Caelomango, L., Huebner, V. D., and Matthews, H. R. (1985) Rapid separation of phosphoamino acids including the phosphohistidines by isocratic high-performance liquid chromatography of the orthophthalaldehyde derivatives. *Anal. Biochem.* **149,** 344–348.

30. McCourt, D. W., Lykam, J. F., and Schwartz, B. D. (1985) Analysis of sulfate and phosphate esters of amino acids by ion-exchange chromatography on polymeric DEAE. *J. Chromatog.* **327,** 9–15.

31. Etheredge, R. W., III, and Glimcher, M. J. (1986) Resolution and identification of O-phosphoserine, O-phosphothreonine, O-phosphotyrosine, and γ-carboxyglutamic acid as their fluorescent *o*-phthalaldehyde derivatives by high-performance liquid chromatography. *Calcif. Tissue Int.* **39,** 239–243.

32. Niedbalski, J. S. and Ringer, D. P. (1986) Separation and quantitative analysis of O-linked phosphoamino acids by isocratic high-performance liquid chromatography of the 9-fluorenylmethyl chloroformate derivatives. *Anal. Biochem.* **158,** 138–145.

33. Ringer, D. P. (1991) Separation of phosphotyrosine, phosphoserine, and phosphothreonine by high-performance liquid chromatography. *Methods Enzymol.* **201,** 3–10.

34. Herber, M., Liedtke, C., Korte, H., Hoffmann-Posorske, E., Donella-Deana, A., Pinna, L. A., et al. (1992) Non-radioactive determination of PTH and dabsyl phosphoamino acids by capillary electrophoresis. *Chromatographia* **33,** 347–350.

35. Meyer, H. E., Eisermann, B., Heber, M., Hoffmann-Posorske, E., Korte, H., Weigt, C., et al. (1993) Strategies for nonradioactive methods in the localization of phosphorylated amino acids in proteins. *FASEB J.* **7,** 776–782.

36. Wang, J. Y. J. (1988) Antibodies for phosphotyrosine: analytical and preparative tool for tyrosyl-phosphorylated proteins. *Anal. Biochem.* **172,** 1–7.
37. Levine, L., Gjika, H. B., and Vunakis, H. V. (1989) Antibodies and radioimmunoassaya for phosphoserine, phosphothreonine and phosphotyrosine. *J. Immunol. Methods* **124,** 239–249.
38. Heffetz, D., Fridkin, M., and Zick, Y. (1989) Antibodies directed against phosphothreonine residues as potent tools for studying protein phosphorylation. *Eur. J. Biochem.* **182,** 343–348.
39. Hunter, T. and Sefton, B. M. (eds.) (1991) Protein phosphorylation part B, in *Methods in Enzynology*, vol. 201, Academic, London and New York.
40. Aitken, A. and Learmonth, M. (1997) Analysis of sites of protein phosphorylation. *Methods Mol. Biol.* **64,** 293–306.
41. Kataoka, H., Sakiyama, N., and Makita, M. (1988) Gas chromatographic analysis of aminoalkylphosphonic acids and aminoalkyl phosphates. *J. Chromatog.* **436,** 67–72.
42. Kataoka, H., Sakiyama, N., and Makita, M. (1990) Occurrence of free O-phosphoserine and O-phosphothreonine in porcine liver. *Agric. Biol. Chem.* **54,** 1731–1733.
43. Kataoka, H., Sakiyama, N., and Makita, M. (1991) Distribution and contents of free O-phosphoamino acids in animal tissues. *J. Biochem.* **109,** 577–580.
44. Kataoka, H., Ueno, Y., and Makita, M. (1991) Analysis of O-phosphoamino acids in proteins by gas chromatography with flame photometric detection. *Agric. Biol. Chem.* **55,** 1587–1592.
45. Kataoka, H., Ueno, Y., and Makita, M. (1992) O-Phosphoamino acid analysis of phosphorylated proteins by gas chromatography with flame photometric detection. *J. Pharm. Biomed. Anal.* **10,** 365–369.
46. Kataoka, H., Nakai, K., Ueno, Y., and Makita, M. (1992) Analysis of O-phosphoamino acids in the protein fractions of mouse tissue by gas chromatography. *Biosci. Biotech. Biochem.* **56,** 1300–1301.
47. Kataoka, H., Nakai, K., and Makita, M. (1993) Identification of O-phosphoamino acids in urine hydrolysate by gas chromatography-mass spectrometry. *J. Chromatog.* **615,** 136–141.
48. Kataoka, H., Nakai, K., Katagiri, Y., and Makita, M. (1993) Analysis of free and bound O-phosphoamino acids in urine by gas chromatography with flame photometric detection. *Biomed. Chromatogr.* **7,** 184–188.
49. Kataoka, H., Nakai, K., and Makita, M. (1994) Increase of phosphotyrosine levels in mouse urine and liver during liver regeneration after partial hepatectomy. *Biochem. Biophys. Res. Commun.* **201,** 909–916.
50. Schlenk, H. and Gellerman, J. L. (1960) Esterification of fatty acids with diazomethane on a small scale. *Anal. Chem.* **32,** 1412–1414.
51. Bidlingmeyer, B. A., Cohen, S. A., and Tarvin, T. L. (1984) Rapid analysis of amino acids using pre-column derivatization. *J. Chromatog.* **336,** 93–104.
52. Makita, M., Yamamoto, S., and Kono, M. (1976) Gas-liquid chromato-graphic analysis of protein amino acids as N-isobutoxycarbonylamino acid methyl esters. *J. Chromatog.* **120,** 129–140.

53. Yamamoto, S., Kiyama, S., Watanabe, Y., and Makita, M. (1982) Practical gas-liquid chromatographic method for the determination of amino acids in human serum. *J. Chromatog.* **233,** 39–50.

54. Martensen, T. M. (1984) Chemical properties, isolation, and analysis of O-phosphates in proteins. *Methods Enzymol.* **107,** 3–23.

55. Bylund, D. B. and Huang, T. S. (1976) Decomposition of phosphoserine and phosphothreonine during acid hydrolysis. *Anal. Biochem.* **73,** 477–485.

15

Determination of Sulfur Amino Acids, Glutathione, and Related Aminothiols in Biological Samples by Gas Chromatography with Flame Photometric Detection

Hiroyuki Kataoka, Kiyomi Takagi, Hirofumi Tanaka, and Masami Makita

1. Introduction

The determination of sulfur amino acids, glutathione (GSH), and related aminothiols has been carried out by isotachophoresis *(1)*, amino acid analyzer (AAA) *(2,3)*, gas chromatography (GC) *(4–8)*, high-performance liquid chromatography (HPLC) *(9–32)*, GC-mass spectrometry (GC-MS) *(33,34)*, liquid chromatography-mass spectrometry (LC-MS) *(35)*, and capillary zone electrophoresis (CZE) *(36–38)*. However, isotachophoresis and AAA methods are nonselective for sulfur amino acids and lack sensitivity. GC methods based on the conversion into trimethylsilyl *(4,5)*, neopentylidine *(6)*, and N-trifluoroacetyl *n*-butyl ester *(7)* derivatives lack sensitivity, gives a tailing peak and requires anhydrous derivatization conditions. Although GC method based on the preparation of N-heptafluorobutyryl isobutyl or ethyl esters *(8)* is selective and sensitive by flame photometric detection (FPD), this method is not applied to the analysis of biological samples. As the HPLC methods, the ultraviolet (UV) *(9–12)*, and the postcolumn UV derivatization with 4,4'-dithiopyridine *(13)* and 5,5-dithiobis (2-nitrobenzoic acid) *(14)*, the precolumn fluorescence derivatization with 4-(aminosulfonyl)- or ammonium-7-fluoro-2,1,3-benzoxadiazole-4-sulfonate *(15,16)*, methyl 4-(6-methoxynaphthalene-2-yl)-4-oxo-2-butenoate *(17)*, N-(1-pyrenyl)maleimide *(18)*, monobromobimane *(19–23)*, *o*-phthaldialdehyde *(24,25)*, 9-fluorenylmethyloxycarbonyl chloride *(26)*, and 2-chloro-1-methylpyridinium *(27)*, and electrochemi-

From: *Methods in Molecular Biology*, vol. 159: *Amino Acid Analysis Protocols*
Edited by: C. Cooper, N. Packer, and K. Williams © Humana Press Inc., Totowa, NJ

cal detection *(28–32)* have been reported. Many of these HPLC methods were highly sensitive, but some of these methods lack specificity and require cleanup of the sample to remove the excess reagent and coexisting substances. GC-MS methods based on the conversion into *tert*-butyldimethylsilyl *(33)* and *N*(O,S)-propoxycarbonyl propyl ester *(34)* derivatives, and LC-MS method were highly sensitive and specific, but these methods require expensive equipment. Furthermore, CZE methods with electrochemical detection are capable of achieving higher separation efficiency, use less organic solvents, and require small amounts of samples in comparison with HPLC, but these methods are not applied enough to the analysis of biological samples. Current methods for the determination of sulfur amino acids, GSH, and related aminothiols have also been described in detail in **refs. *39–43***.

Recently, we have developed a selective and sensitive method for the determination of sulfur amino acids by FPD-GC using a DB-17 capillary column (15 m × 0.53-mm id), in which these compounds were analyzed as their *N*(S)-isopropoxycarbonyl (isoPOC) methyl ester derivatives *(44)*. By using this method, we demonstrated that the contents of these amino acids in urine *(44)* and plasma *(45)* samples could be rapidly and simply determined. Furthermore, we have developed a selective and sensitive method for the determination of GSH and related aminothiols such as cysteine (Cys), cysteinylglycine (CysGly), and γ-glutamylcysteine (γ-GluCys) by FPD-GC, using a short capillary column (5 m × 0.53-mm id) of crosslinked DB-1 *(46)*. By using this method, we demonstrated that the contents of these aminothiols in blood *(46)* and tissue *(47)* samples could be rapidly and simply determined. In the aforementioned methods, other sulfur containing amino acids such as taurine and related aminosulfonic acids could not be analyzed, but these compounds could be analyzed as their *N*-isoBOC di-*n*-butylamide derivatives by GC *(48–52)*. In this chapter, selective and sensitive method for the determination of sulfur amino acids, GSH, and related aminothiols, except for aminosulfonic acids by FPD-GC, is described on the basis of the above results, and optimum conditions and typical problems encountered in the development and application of the method are discussed.

The derivatization process of sulfur amino acids, GSH, and related aminothiols and structures of main compounds are shown in **Fig. 1**. *N*(S)-isoPOC methyl esters of these compounds can be easily, rapidly, and quantitatively prepared by the reaction with isopropyl chloroformate (isoPCF) by shaking in aqueous alkaline media, followed by esterification with hydrogen chloride-methanol (HCl-MeOH). The oxidized and protein-bound forms of these compounds were reduced to free thiols by adding sodium borohydride (NaBH$_4$) prior to derivatization. The main advantage of this method is that these amino acids can be easily converted in an aqueous medium without any

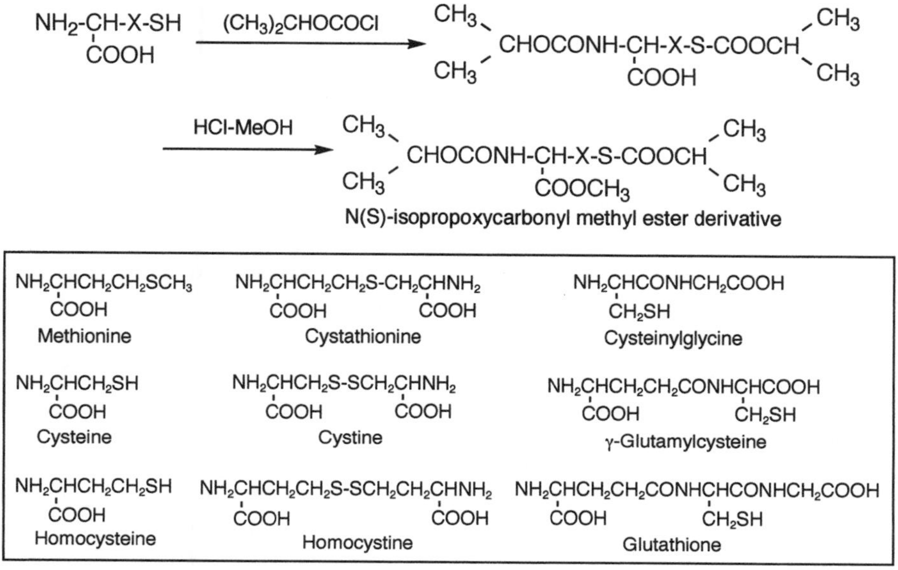

NH₂-CH-X-SH (CH₃)₂CHOCOCl ——→ CH₃╲CHOCONH-CH-X-S-COOCH╱CH₃
  | CH₃╱ | ╲CH₃
  COOH   COOH

$$\text{NH}_2\text{-CH-X-SH} \quad \xrightarrow{(\text{CH}_3)_2\text{CHOCOCl}} \quad$$

———HCl-MeOH——→ CH₃╲CHOCONH-CH-X-S-COOCH╱CH₃
                 CH₃╱ | ╲CH₃
                          COOCH₃

N(S)-isopropoxycarbonyl methyl ester derivative

NH₂CHCH₂CH₂SCH₃
  COOH
Methionine
	NH₂CHCH₂CH₂S-CH₂CHNH₂
	  COOH COOH
	Cystathionine
		NH₂CHCONHCH₂COOH
		  CH₂SH
		Cysteinylglycine

NH₂CHCH₂SH
  COOH
Cysteine
	NH₂CHCH₂S-SCH₂CHNH₂
	  COOH COOH
	Cystine
		NH₂CHCH₂CH₂CONHCHCOOH
		  COOH CH₂SH
		γ-Glutamylcysteine

NH₂CHCH₂CH₂SH
  COOH
Homocysteine
	NH₂CHCH₂CH₂S-SCH₂CH₂CHNH₂
	  COOH COOH
	Homocystine
		NH₂CHCH₂CH₂CONHCHCONHCH₂COOH
		  COOH CH₂SH
		Glutathione

Fig. 1. Derivatization process and structures of sulfur amino acids, glutathione, and related aminothiols.

further clean-up procedure and the derivatives are stable to moisture. Furthermore, the derivatives can be quantitatively and reproducibly resolved as single peaks using a capillary column, and provide an excellent FPD response.

2. Materials

2.1. Equipment

1. Shimadzu 12A gas chromatograph equipped with a hydrogen flame ionization detector and a flame photometric detector (S-filter).
2. For the analysis of sulfur amino acids, a fused-silica capillary column (15 m × 0.53-mm id, 1.0-μm film thickness) of crosslinked DB-17 (50% phenyl-50% methylpolysiloxane, J & W, Folsom, CA).
3. For the analysis of GSH and related aminothiols, a fused-silica capillary column (5 m × 0.53-mm id, 1.5-μm film thickness) of crosslinked DB-1 (methylpolysiloxane, J & W).
4. Model LK–21 ultradisperser (Yamato Kagaku, Tokyo, Japan).
5. Pasteur capillary pipet (Iwaki glass no. IK-PAS-5P).

2.2. Reagents

1. Amino acids and peptides: L-methionine (Met), ethionine (Eth), L-thioproline (TPro), L-cysteine (Cys), D,L-homocysteine (HCys), L-cystine (Cyt), D,L-ho-

mocystine (HCyt), and djenkolic acid (Dje) were purchased from Nacalai Tesque (Kyoto, Japan). L-Cystathionine (CTH), L-methionine sulfone (Met-S), S-methyl-L-cysteine (SM-Cys), S-carboxymethyl L-cysteine (SC-Cys) and cysteinylglycine (CysGly) were purchased from Sigma (St. Louis, MO). Glutathione (GSH) and γ-glutamylcysteine (γ-GluCys) were purchased from Kohjin Co., Ltd. (Tokyo, Japan). Glutathione disulphide (GSSG) was purchased from Wako Pure Chemical Industries (Osaka, Japan). All of the nonsulfur protein amino acids used were purchased from Ajinomoto (Tokyo, Japan).

2. Internal standards: S-2-Aminoethyl-L-cysteine (AE-Cys) and lanthionine (LTH) as internal standard (IS) were purchased from Sigma.

3. Derivatizing reagents: isopropyl chloroformate (isoPCF) was obtained from Wako Pure Chemicals Industries (Osaka, Japan). Hydrogen chloride in methanol (HCl-MeOH) obtained from Tokyo Kasei Kogyo (Tokyo, Japan) was diluted with methanol at a concentration of 1 M.

4. Other materials: Sodium borohydride (NaBH$_4$) and dithioerythritol (DTE) were obtained from Nacalai Tesque. Peroxide-free diethyl ether was obtained from Dojindo Laboratories (Kumamoto, Japan). Distilled water was used after freshly purification with a Model Milli-Q Jr. water purifier (Millipore, Bedford, MA). All other chemicals were of analytical-reagent grade.

2.3. Solutions

1. Standard sulfur amino acid solutions: three standard stock solutions (each 2 m*M*), one containing the 13 sulfur amino acids except for GSH, CysGly, and γ-GluCys, the second containing Cys, CysGly, and γ-GluCys, and the third containing nonsulfur protein amino acids; are prepared in 0.05 *M* HCl.

2. Working standard solutions: each working standard solution is made up freshly as required by dilution of the stock solution with 0.01 *M* HCl. When these standard solutions are stored at 4°C, they are stable for at least 2 wk.

3. The GSH standard solution: GSH is freshly dissolved in distilled water at the required concentration on each day, because GSH is partialy hydrolyzed to γ-GluCys during storage.

4. The internal standard solutions: AE-Cys and LTH are dissolved in 0.05 *M* HCl to make a stock solution at a concentration of 2 m*M*. Store at 4°C.

5. NaBH$_4$ solution: NaBH$_4$ is dissolved in 0.1 *M* NaOH at a concentration of 100 mg/mL solution. This solution is stable at 4°C for at least 1 wk.

6. DTE solution: DTE is dissolved in distilled water at a concentration of 0.5 m*M*.

3. Methods

3.1. Derivatization

Derivatization procedure of sulfur amino acids, GSH and related aminothiols is shown in **Fig. 2** (*see* **Note 1**).

1. For the analysis of 13 sulfur amino acids, pipet an aliquot of the sample containing 0.5–10 nmol of each amino acid and 0.1 mL of 2 μ*M* AE-Cys (IS) (if necessary) into a 10-mL Pyrex glass reaction tube with a PTFE-lined screw cap.

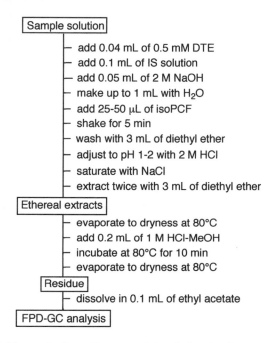

Fig. 2. Shematic flow diagram of the derivatization procedure.

2. For the analysis of GSH and related aminothiols, pipet an aliquot of the sample containing 0.2–25 nmol of each compounds and 0.1 mL of 10 μ*M* LTH (IS) (if necessary) into a reaction tube.

3. To these solutions, add 40 μL of 0.5 m*M* DTE (*see* **Note 2**) and 0.05 mL of 2 *M* NaOH, and make the total volume up to 1 mL with distilled water.

4. Immediately add 25–50 μL of isoPCF and shake the mixture at 300 cycles per minute (up and down) for 5 min at room temperature (*see* **Note 3**).

5. Extract the reaction mixture with 3 mL of peroxide-free diethyl ether (*see* **Note 2**), to remove the excess of reagent, with vigorous shaking for 5–10 s by hand (*see* **Note 4**).

6. Centrifuge at 2000*g* for 30 s then discard the etheral extract.

7. Acidify the aqueous layer to pH 1–2 with 2 *M* HCl and saturate with NaCl.

8. Extract the mixture twice with 3 mL of peroxide-free diethyl ether with vigorous shaking for 5–10 s by hand (*see* **Note 5**).

9. Repeat centrifugation at 2000*g* for 30 s, and transfer the ether layers into another tube by means of a Pasteur capillary pipet.

10. Evaporate the ethereal extracts to dryness at 80°C, and to the residue add 0.2 mL of 1 *M* HCl-MeOH. Incubate the mixture at 80°C for 10 min (*see* **Note 6**).

11. Evaporate the residual solvent to dryness at 80°C under a stream of dry air. Re-dissolve the residue on the walls of the tube in 0.1–0.2 mL of ethyl acetate and inject 0.2–1 μL of this solution into the gas chromatograph (*see* **Note 7**).

3.2. Gas Chromatography

1. GC analysis was carried out with a Shimadzu 12A gas chromatograph equipped with a flame ionization detector and a flame photometric detector (S-filter) (*see* **Note 8**).
2. For the analysis of sulfur amino acids, a fused-silica capillary column (15 m × 0.53-mm id, 1.0-μm film thickness) of crosslinked DB-17 is used. The operating conditions are as follows: column temperature program at 5°C/min from 130°C to 270°C or at 10°C/min from 150°C to 270°C; injection and detector temperature, 280°C; nitrogen flow rate, 10 mL/min (*see* **Note 9**).
3. For the analysis of GSH and related aminothiols, a fused-silica capillary column (5 m × 0.53-mm id, 1.5-μm film thickness) of crosslinked DB-1 is used. The operating conditions are as follows: column temperature program at 16°C/min from 130°C to 290°C; injection and detector temperature, 300°C; nitrogen flow rate, 12 mL/min (*see* **Notes 10** and **11**).
4. Measure the peak heights of each compound and the IS. Calculate the peak height ratios against the IS to construct calibration curves (*see* **Notes 12–15**).

3.3. Sample Preparation *(see **Note 16**)*

3.3.1. Preparation of Urine Samples *(44)*

1. Collect early morning urine samples from healthy volunteers and processed immediately or store at –20°C until used.
2. Urine samples (0.02–0.1 mL) are used directly for derivatization and FPD-GC analysis (*see* **Note 17**).

3.3.2. Preparation of Blood Samples *(46)*

1. Collect venous blood samples from healthy volunteers in 9 vol of 1 mM ethylenediamine tetraacetate (EDTA) and mix (*see* **Note 18**).
2. Immediately process the 10% (v/v) clear hemolysates or store at –20°C until used.
3. To determine the free GSH, Cys, CysGly, and γ-GluCys contents, 0.1 mL of 10% hemolysate is directly used as the sample for derivatization.
4. To obtain total (free and bound, reduced, and oxidized) GSH, Cys, CysGly, and γ-GluCys, chemically reduce a blood sample with NaBH$_4$ (*see* **Note 19**). To 0.05 mL of 10% hemolysate add 0.1 mL of 10 μM LTH (IS) and 0.2 mL of 100 mg/mL NaBH$_4$, and then make the total volume up to 1 mL with distilled water. After adding 1 drop of *n*-hexanol (*see* **Note 20**), the mixture is incubated at 100°C for 10 min (*see* **Note 21**). After cooling, the reaction mixture is used as the sample for derivatization.

3.3.3. Preparation of Tissue Samples *(47)*

The outline of the method for determining the different forms of GSH and related aminothiols in tissue samples is shown in **Fig. 3** (*see* **Note 22**). Five male ddY mice 6 wk old (28.0 ± 0.9 g) were used in the experiments.

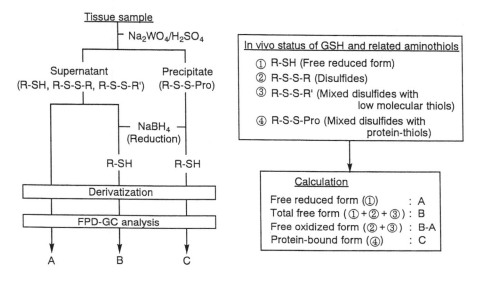

Fig. 3. Outline of method for the determination of different forms of glutathione and related aminothiols in tissue samples.

1. Immediately after dissection, each organ should be removed and stored at –20°C until used.
2. Homogenize an aliquot of each pooled tissue (0.2–0.4 g) with 2.4 mL of 0.05 M H_2SO_4 and 1.2 mL of 2.5% Na_2WO_4 with an LK–21 ultradisperser (Yamato Kagaku, Tokyo, Japan) after adding one drop of n-hexanol.
3. Centrifuge at 2000g for 5 min, then reextract the precipitate with the same volume of 0.05 M H_2SO_4 and 2.5% Na_2WO_4 as above.
4. Combine the supernatants and make up to 10 mL with distilled water. This solution is then ready to use to determine the free and free oxidized forms.
5. For the analysis of free GSH and other aminothiols, 0.1 mL of a free fraction is directly derivatized.
6. For the analysis of total free GSH and other aminothiols (thiols + disulfides + mixed disulfides with other low-molecular-weight thiols), add 0.1 mL of the free fraction to 0.2 mL of 100 mg/mL of $NaBH_4$ in a 0.01 M NaOH solution, and incubate the mixture at 60°C for 5 min. After cooling, the reaction mixture is used as the sample for derivatization.
7. On the other hand, the precipitate is dissolved in 5 mL of 0.01 M NaOH and made up to 10 mL with distilled water, before being used to determine the protein-bound form. For the analysis of protein-bound GSH and other aminothiols, add 0.5 mL of the protein-bound fraction to 0.2 mL of 100 mg/mL of $NaBH_4$ in a 0.01 M NaOH solution, and incubate the mixture at 60°C for 5 min. After cooling, the reaction mixture is used as the sample for derivatization.

4. Notes

1. IsoPCF and HCl-MeOH were used as derivatizing reagents for the amino and sulfhydryl groups, and for the carboxyl group, respectively (**Fig. 1**).

2. In order to prevent the oxidation of sulfur function during derivatization, DTE was added to reaction mixture and peroxide-free diethyl ether was used as an extraction solvent.

3. The *N*(S)-isopropoxycarbonylation of sulfur amino acids, GSH, and related aminothiols with isoPCF proceeded rapidly and quantitatively in aqueous alkaline media *(44,46)*. This reaction was completed within 5 min by shaking at room temperature.

4. The reaction mixture was then washed with diethyl ether under alkaline condition in order to remove the excess reagent. This procedure also serves to exclude amines and phenols both of which often coexist with amino acids in biological samples, as they are derivatized to the corresponding N- or O-isoPOC derivatives, which are soluble in organic solvents under the same conditions as aforementioned.

5. The resulting *N*(S)-isoPOC derivatives in aqueous layer were quantitatively and selectively extracted into diethyl ether after acidification to pH 1.0–2.0. In this procedure, the ether layers should be collected, taking care to avoid aqueous droplets. It was not necessary to complete draw the ether layer in each extraction.

6. Although the methylation of the carboxyl group with diazomethane was simple *(53,54)*, the interfering peaks originating in the diazomethane generating reagents were observed under our FPD-GC conditions. The methylation of *N*(S)-isoPOC derivatives with HCl-MeOH was completed within 5 min at 80°C, and no interfering peak was observed.

7. The derivative preparation was accomplished within 30 min, and several samples could be treated simultaneously. The *N*(S)-isoPOC methyl ester derivatives of sulfur amino acids and related compounds were stable under nomal laboratory conditions and no decomposition was observed during GC analysis.

8. The GC analysis was performed with FPD (394-nm interference filter), which is highly selective for sulfur containing compounds, by using a megabore capillary column and direct (splitless) sample injection system. In this system, the sufficiently inactivated glass insert should be used to avoid tailing peaks of derivatives caused by the adsorption on injection port.

9. Nonsulfur amino acids were also derivatized and detected with flame ionization detection (FID) (**Fig. 4A**), but these amino acids were not detected at all with FPD (**Fig. 4C**). On the other hand, sulfur amino acids were detected with both FID (**Fig. 4B**) and FPD (**Fig. 4D**). As shown in **Fig. 4**, 13 sulfur amino acids and IS could be completely resolved as single and symmeterical peaks within 32 min on a DB-17 capillary column (15 m × 0.53-mm id, 1.0-μm film thickness) *(44)*.

10. The separation of biologically important sulfur amino acids was achieved within 18 min on a DB-17 capillary column (**Fig. 5A**). However, the derivatives of GSH and γ-GluCys could not be eluted with this capillary column because of high molecular mass and high boiling point of these derivatives.

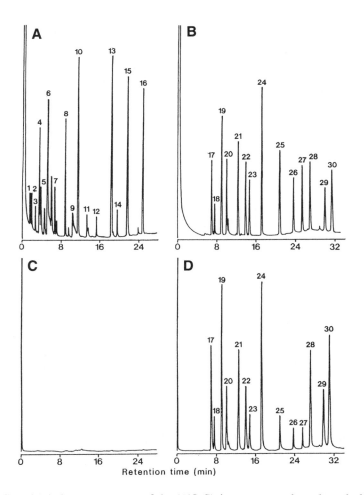

Fig. 4. Standard chromatograms of the N(O,S)-isoprpoxycarbonyl methyl ester derivatives of nonsulfur and sulfur amino acids. (A) Nonsulfur amino acids (containing 200 nmol of each amino acid); (B) sulfur amino acids (containing 200 nmol of each amino acid); (C) nonsulfur amino acids (containing 50 nmol of each amino acid); (D) sulfur amino acids (containing 1 nmol of each amino acid). The derivatized samples were analyzed by FID-GC (A and B) and FPD-GC (C and D). GC conditions: column, DB-17 (15 m × 0.53-mm id, 1.0-μm film thickness); column temperature, programmed at 5°C/min from 130°C to 270°C and then held for 5 min; injection and detector temperatures, 280°C; nitrogen carrier gas flow rate, 10 mL/min. Peaks: 1 = alanine; 2 = glycine; 3 = valine; 4 = leucine; 5 = isoleucine; 6 = serine + threonine + proline; 7 = aspartic acid; 8 = glutamic acid; 9 = hydroxyproline; 10 = phenylalanine; 11 = asparagine; 12 = glutamine; 13 = lysine; 14 = histidine; 15 = tyrosine; 16 = tryptophan; 17 = S-methylcysteine (SM-Cys); 18 = thioproline (TPro); 19 = methionine (Met); 20 = ethionine (Eth); 21 = cysteine (Cys); 22 = S-carboxymethylcysteine (SC-Cys); 23 = homocysteine (HCys); 24 = methionine sulfone (Met-S); 25 = S-2-aminoethylcysteine (AE-Cys) (IS); 26 = lanthionine (LTH); 27 = cystathionine (CTH); 28 = cystine (Cyt); 29 = djenkolic acid (Dje); 30 = homocystine (HCyt) (from **ref. 44**, with permission).

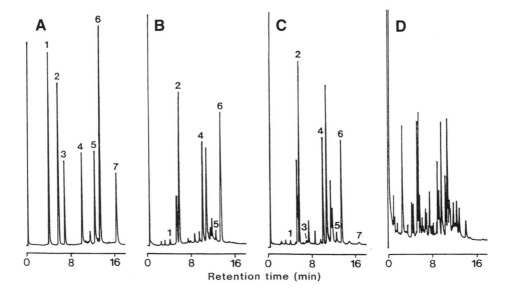

Fig. 5. Chromatograms obtained from standard and urine samples. (**A**) Standard (containing 2 nmol of each amino acid); (**B**) urine A (20 μL); (**C**) urine B (20 μL); (**D**) urine B (0.8 mL). The derivatized samples were analyzed by FPD-GC (**A–C**) and FID-GC (**D**). GC conditions: column, DB-17 (15 m × 0.53-mm id, 1.0-μm film thickness); column temperature, programmed at 10°C/min from 150°C to 270°C and then held for 5 min; injection and detector temperatures, 280°C; nitrogen carrier gas flow rate, 10 mL/min. Peaks: 1 = methionine (Met); 2 = cysteine (Cys); 3 = homocysteine (HCys); 4 = S-2-aminoethylcysteine (AE-Cys) (IS); 5 = cystathionine (CTH); 6 = cystine (Cyt); 7 = homocystine (HCyt) (from **ref. 44**, with permission).

11. In order to solve above problem, a column was cut short. Of different length of several columns tested, DB-1 (5 m × 0.53-mm id, 1.5-μm film thickness) proved to be the most satisfactory column for the purpose. By using this column, Cys, CysGly, γ-GluCys, GSH, and IS were separately eluted as single and symmetrical peaks within 10 min (**Fig. 6A**) *(46)*. However, GSSG, disulfide form of GSH, could not be eluted even with this column.

12. The calibration curves for sulfur amino acids by FPD-GC were conducted using AE-Cys, which showed a similar behavior to other sulfur amino acids during the derivatization and was well separated from these amino acids on a chromatogram as the IS *(44,45)*. A linear relationship was obtained from double logarithmic plots in the range 0.5–10 nmol of sulfur amino acids, and correlation coefficients were above 0.998 (**Table 1**).

13. The derivatives of sulfur amino acids provided excellent FPD responses and the minimum detectable amounts of these amino acids at a signal three times as high as the noise under our instrumental conditions were approx 0.5–1.0 pmol as injec-

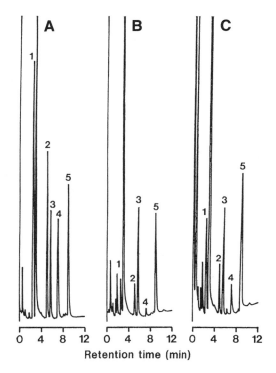

Fig. 6. Chromatograms obtained from standard and whole blood samples. (**A**) Standard (containing 5 nmol of glutathione and 1 nmol of other aminothiols); (**B**) blood (10 mL, nonreduction); (C) blood (5 mL, reduction). GC conditions: column, DB-1 (5 m × 0.53-mm id, 1.5-μm film thickness); column temperature, programmed at 16°C/min from 130°C to 290°C; injection and detection temperature, 300°C; nitrogen flow rate, 12 mL/min. Peaks: 1 = cysteine (Cys), 2 = cysteinylglycine (CysGly), 3 = lanthionine (LTH) (IS), 4 = γ-glutamylcysteine (γ-GluCys), 5 = glutathione (GSH) (from **ref. *46***, with permission).

tion amounts. The FPD-GC system described here was over 20 times more sensitive than the FID-GC system.

14. The calibration curves for GSH and related aminothiols by FPD-GC were conducted using LTH, which showed a similar behavior to other aminothiols during the derivatization and was well separated from these aminothiols on a chromatogram as the IS (*46,47*). A linear relationship was obtained from double-logarithmic plots in the range 1–25 nmol for GSH and 0.2–5 nmol for other aminothiols, and correlation coefficients were above 0.999 (**Table 2**).

15. The minimum detectable amount of GHS and related aminothiols to give a signal three times as high as the noise under our FPD-GC conditions were approx 0.2–5 pmol as injection amounts.

Table 1
Linear Regression, Detection Limits, and Recoveries from Urine Samples on the Determination of Sulfur Amino Acids

Sulfur amino acid	Regression line[a] Slope	Intercept	Correlation coefficient	Detection limit (pmol)	Urine A Amount found (nmol/mL)[b] Non-addition	Addition[c]	Recovery (%)	Urine B Amount found (nmol/mL)[b] Non-addition	Addition[c]	Recovery (%)
Met	1.822	−0.222	0.9983	0.5	12.9 ± 0.6	111.0 ± 6.2	98.1	8.0 ± 0.5	93.0 ± 0.3	85.0
Cys	1.777	−0.308	0.9989	0.5	33.9 ± 1.2	136.5 ± 1.3	102.6	69.0 ± 1.5	165.0 ± 4.5	96.0
HCys	1.703	−0.514	0.9995	0.5	ND[d]	90.0 ± 2.2	90.9	ND	92.5 ± 2.5	92.5
CTH	1.781	−0.548	0.9998	1.0	22.1 ± 0.9	122.0 ± 1.2	99.9	22.0 ± 0.3	114.5 ± 2.5	92.5
Cyt	1.834	−0.188	0.9998	1.0	45.0 ± 1.2	145.0 ± 0.9	100.0	61.5 ± 0.3	159.0 ± 3.5	97.5
HCyt	1.859	−0.059	0.9986	1.0	ND	113.0 ± 1.3	113.0	ND	107.5 ± 1.5	107.5

[a]$\log y = a \log x + b$: y, peak height ratio against the IS; x, amount of each amino acid (nmol); a, slope; b, intercept. Range: 0.5–10 nmol.
[b]Mean ± SD ($n = 3$).
[c]Addition: 100 nmol/mL.
[d]Not detectable.

Table 2
Linear Regression, Detection Limits, and Recoveries from Blood Samples on the Determination of GSH and Related Aminothiols

Aminothiol	Regression line[a]			Detection limit (pmol)	Non-reduced blood			Reduced blood		
	Slope	Intercept	Correlation coefficient		Amount found (nmol/mL)[b]		Recovery (%)	Amount found (nmol/mL)[b]		Recovery (%)
					Non-addition	Addition[c]		Non-addition	Addition[c]	
Cys	1.418	0.367	0.9997	0.2	19.0 ± 1.4	111.2 ± 3.8	92.2	83.3 ± 2.2	260.7 ± 8.3	88.7
CysGly	1.419	0.186	0.9992	0.3	19.5 ± 0.9	110.2 ± 8.3	90.7	55.4 ± 2.6	248.9 ± 13.1	96.8
γ-GluCys	1.419	-0.029	0.9990	0.5	14.3 ± 1.0	104.6 ± 3.2	90.3	72.6 ± 4.8	286.0 ± 7.7	106.7
GSH	1.677	-1.154	0.9993	5.0	464 ± 34	954.0 ± 42	98.0	958 ± 55	1977 ± 121	101.9

[a]$\log y = a \log x + b : y$, peak height ratio against the IS; x, amount of each amino acid (nmol); a, slope; b, intercept. Range: 1–25 nmol for GSH, 0.2–5 nmol for the other aminothiols.

[b]Mean ± SD ($n = 3$).

[c]Addition: 500 nmol/mL for GSH, 100 nmol/mL for the other aminothiols.

[d]Addition: 1000 nmol/mL for GSH, 200 nmol/mL for the other aminothiols.

16. The method developed for sulfur amino acids was successfully applied to small urine (0.02–0.1 mL) *(44)* and plasma (0.1 mL) *(45)* samples without prior clean-up of the sample. As shown in **Fig. 5B,C**, unknown peaks were observed in front of the Cys peak and in the near of the IS peak, but Met, Cys, CTH, Cyt, and trace quantities of HCys and HCyt were detected without any interference from coexisting substances when the sample was analyzed by FPD-GC.

17. It was difficult to determine sulfur amino acids in urine by FID-GC (**Fig. 5D**). As shown in **Table 1**, the overall recoveries of sulfur amino acids added to urine samples were 85–113% and the relative standard deviations were 0.3–6.4% ($n = 3$).

18. The method developed for GSH and related aminothiols was successfully applied to blood sample, in which whole blood was used because it should reflect more accurately the total GSH status of blood including GSH bound to plasma proteins, and erythrocytes contain almost all of the GSH *(55)*. Hemolysis of red blood cell was carried out in solution of EDTA to prevent oxidation of thiol function by any metallic elements. The hemolysate of small blood (5–10 µL) could be used without prior clean-up of the sample *(46)*. **Figure 6B,C** show typical chromatograms obtained from nonreduced and reduced blood samples, respectively. Free and total GSH and related aminothiols in whole blood sample could be analyzed without any interference from coexisting substances. As shown in **Table 2**, the overall recoveries of GSH and related aminothiols added to whole blood samples were 88–107% and the relative standard deviations were 2.6–7.5% ($n = 4$).

19. In order to determine the redox status of sulfur amino acids *(45)*, GSH, and related aminothiols *(46,47)*, disulfides in the sample must be reduced to the corresponding thiols prior to derivatization. NaBH$_4$ efficiently converts all these aminothiols present as free thiols and mixed disulfides with other low-molecular mass thiols or protein-thiols to thiol forms *(20–22)*.

20. Formation of gas and foaming during the NaBH$_4$ reduction of biological samples could be reduced by adding one drop of *n*-hexanol, surface active agent. The reduced sample could be directly derivatized after cooling.

21. In order to confirm the reduction conditions, some disulfide compounds such as Cyt, HCyt, and GSSG were reduced with NaBH$_4$. The reduction of these compounds were accomplished within 10 min at 100°C by using 20 mg NaBH$_4$ in aqueous alkaline medium.

22. In order to determine the concentration of the different forms of GSH and related aminothiols in tissue samples, the fractionation of the tissue samples was investigated according to the previous method *(56)*. As shown in **Fig. 3**, the sample was separated into its free and protein-bound fractions by precipitation with Na$_2$WO$_4$-H$_2$SO$_4$. No free GSH or other aminothiols were apparent in the protein-bound fraction. **Figure 7B–D** shows typical chromatograms obtained from the free fraction (-reduction), free fraction (+reduction), and protein-bound fraction (+reduction), respectively. Each form of GSH and related aminothiols in a tissue sample could be analyzed without any interference from coexisting substances. The overall recoveries of GSH and related aminothiols added to the free and protein-bound

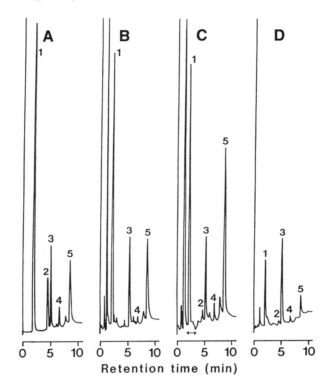

Fig. 7. Gas chromatograms obtained from standard solution and free and protein-bound fractions of mouse liver. (**A**) Standard containing 0.25 nmol of cysteine, 0.5 nmol of cysteinylglycine and γ-glutamylcysteine and 5 nmol of glutathione; (**B**) free fraction (–reduction); (**C**) free fraction (+reduction); (**D**) protein-bound fraction (+reduction). GC conditions and peak no.: *see* **Fig. 6**. The double arrow on the time-scale indicates the time during which the recorder response was reduced to half (from **ref. 47** with permission).

fractions of mouse liver were 91.2–109.6%, and the relative standard deviation was 1.1–10.7% ($n = 4$). Oxidized GSH and related aminothiols (disulfides and mixed disulfides with low-molecular weight thiols) were measured by subtracting their free reduced contents from their total free contents.

References

1. Mizobuchi, N., Ageta, T., Sasaki, K., and Kodama, H. (1986) Isotacho- phoretic analysis of cystine, homocystine and cystathionine in urines from patients with inborn errors of metabolism. *J. Chromatog.* **382,** 321–325.
2. Friedman, M., Noma, A. T., and Wagner, J. R. (1979) Ion-exchange chromatography of sulfur amino acids on a single-column amino acid analyzer. *Anal. Biochem.* **98,** 293–304.

3. Andersson, A., Brattstrom, L., Isaksson, A., Israelsson, B., and Hultberg, B. (1989) Determination of homocysteine in plasma by ion-exchange chromatography. *J. Clin. Lab. Invest.* **49,** 445–449.

4. Caldwell, K. A. and Tappel, A. L. (1968) Separation by gas-liquid chromatography of silylated derivatives of some sulfo- and selenoamino acids and their oxidation products. *J. Chromatog.* **32,** 635–640.

5. Shahrokhi, F. and Gehrke, C. W. (1968) Quantitative gas-liquid chromato- graphy of sulfur containing amino acids. *J. Chromatog.* **36,** 31–41.

6. Jellum, E., Bacon, V. A., Patton, W., Pereira, W., Jr., and Halpern, B. (1969) Quantitative determination of biologically important thiols and disulfides by gas-liquid chromatography. *Anal. Biochem.* **31,** 339–347.

7. Mee, J. M. L. (1973) Gas-liquid chromatographic analysis of glutathione. *J. Chromatog.* **87,** 258–262.

8. Bonvel, S. I. and Monheimer, R. H. (1980) A gas chromatographic analysis of sulfur-containing amino acids employing flame photometric detection. *J. Chromatog. Sci.* **18,** 18–22.

9. Reed, D. J., Babson, J. R., Beatty, P. W., Brodie, A. E., Ellis, W. W., and Potter, D. W. (1980) High-performance liquid chromatography analysis of nanomole levels of glutathione, glutathione disulfide, and related thiols and disulfides. *Anal. Biochem.* **106,** 55–62.

10. Di Pietra, A. M., Gotti, R., Bonazzi, D., Andrisano, V., and Cavrini, V. (1994) HPLC determination of glutathione and L-cysteine in pharmaceuticals after derivatization with ethacrynic acid. *J. Pharm. Biomed. Anal.* **12,** 91–98.

11. Asensi, M., Sastre, J., Pallardo, F. V., de la Asuncion, J. G., Estrela, J. M., and Vina, J. (1994) A high-performance liquid chromatography method for measurement of oxidized glutathione in biological samples. *Anal. Biochem.* **217,** 323–328.

12. Yoshida, T. (1996) Determination of reduced and oxidized glutathione in erythrocytes by high-performance liquid chromatography with ultraviolet absorbance detection. *J. Chromatog. B* **678,** 157–164.

13. Andersson, A., Isaksson, A., Brattstrom, L., and Hultberg, B. (1993) Homocysteine and other thiols determined in plasma by HPLC and thiol-specific postcolumn derivatization. *Clin. Chem.* **39,** 1590–1597.

14. Nozal, M. J., Bernal, J. L., Toribio, L., Marinero, P., Moral, O., Manzanas, L., and Rodriguez, E. (1997) Determination of glutathione, cysteine and N-acetylcysteine in rabitt eye tissues using high-performance liquid chromatography and post-column derivatization with 5, 5-dithiobis(2-nitrobenzoic acid). *J. Chromatog. A,* **778,** 347–353.

15. Ubbink, J. B., Vermaak, W. J. H., and Bissbort, S. (1991) Rapid high- performance liquid chromatographic assay for total homocysteine levels in human serum. *J. Chromatog.* **565,** 441–446.

16. Cornwell, P. E., Morgan, S. L., and Vaughn, W. H. (1993) Modification of a high-performance liquid chromatographic method for assay of homocycteine in human plasma. *J. Chromatog.* **617,** 136–139.

17. Gotti, R., Andrisano, V., Gotti, R., Cavrini, V., and Candeletti, S. (1994) Determination of glutathione in biological samples by high performance liquid chromatography with fluorescence detection. *Biomed. Chromatogr.* **8,** 306–308.
18. Winters, R. A., Zukowski, J., Ercal, N., Mattews, R. H., and Spitz, D. R. (1995) Analysis of glutathione, glutathione disulfide, cysteine, homocysteine, and other biological thiols by high-performance liquid chromatography following derivatization by N-(1-pyrenyl)maleimide. *Anal. Biochem.* **227,** 14–21.
19. Newton, G. L., Dorian, R., and Fahey, R. C. (1981) Analysis of biological thiols: Derivatization with monobromobimane and separation by reversed-phase high-performance liquid chromatography. *Anal. Biochem.* **114,** 383–387.
20. Jacobsen, D. W., Gatautis, V. J., and Green, R. (1989) Determination of plasma homocysteine by high-performance liquid chromatography with fluorescence detection. *Anal. Biochem.* **178,** 208–214.
21. Mansoor, M. A., Svardal, A. M., and Ueland, P. M. (1992) Determination of the in vivo redox status of cysteine, cysteinylglycine, homocysteine, and glutathione in human plasma. *Anal. Biochem.* **200,** 218–229.
22. Fiskerstrand, T., Refsum, H., Kvalheim, G., and Ueland, P. M. (1993) Honocysteine and other thiols in plasma and urine: Automated determination and sample stability. *Clin. Chem.* **39,** 263–271.
23. Yan, C. C. and Huxtable, R. J. (1995) Fluorimetric determination of monobromobimane and o-phthalaldehyde adducts of γ-glutamylcysteine and glutathione: application to assay of γ-glutamylcysteine synthetase activity and glutathione concentration in liver. *J. Chromatog. B* **672,** 217–224.
24. Fermo, I., Arcelloni, C., De Vecchi, E., Vigano, S., and Paroni, R. (1992) High-performance liquid chromatographic method with fluorescence detection for the determination of total homocyst(e)ine in plasma. *J. Chromatog.* **593,** 171–176.
25. Paroni, R., DeVecchi, E., Cighetli, G., Arecelloni, C., Fermo, I., Grossi, A., and Bonini, P. (1995) HPLC with o-phthalaldehyde precolumn derivatization to measure total, oxidized, and protein-bound glutathione in blood, plasma, and tissue. *Clin. Chem.* **41,** 448–454.
26. Staffeldt, B., Brockmoller, J., and Roots, I. (1991) Determination of S- carboxymethyl-L-cysteine and some of its metabolites in urine and serum by high-performance liquid chromatography using fluorescence pre-column labelling. *J. Chromatog.* **571,** 133–147.
27. Sypniewski, S. and Bald, E. (1994) Ion-pair high-performance liquid chromatography of cysteine and metabolically related compounds in the form of their S-pyridinium derivatives. *J. Chromatog. A* **676,** 321–330.
28. Richie, J. P. and Lang, C. A. (1987) The determination of glutathione, cyst(e)ine, and other thiols and disulfides in biological samples using high-performance liquid chromatography with dual electrochemical detection. *Anal. Biochem.* **163,** 9–15.
29. Carvalho, F. D., Remiao, F., Vale, P., Timbrell, J. A., Bastos, M. L., and Ferreira, M. A. (1994) Glutathione and cysteine measurement in biological samples by HPLC with a glassy carbon working detector. *Biomed. Chromatogr.* **8,** 134–136.

30. Yang, C.-S., Tsai, P.-J., Chen, W.-Y., Liu, L., and Kuo, J.-S. (1995) Determination of extracellular glutathione in livers of anaesthetized rats by microdialysis with on-line high-performance liquid chromatography. *J. Chromatog.* **667,** 41–48.

31. Kleinman, W. A. and Richie, J. P., Jr. (1995) Determination of thiols and disulfides using high-performance liquid chromatography with electrochemical detection. *J. Chromatog. B* **672,** 73–80.

32. Lakritz, J., Plopper, C. G., and Buckpitt, A. R. (1997) Validated high-performance liquid chromatography-electrochemical method for determination of glutathione and glutathione disulfide in small tissue samples. *Anal. Biochem.* **247,** 63–68.

33. Stabler, S. P., Marcell, P. D., Podell, E. R., and Allen, R. H. (1987) Quantitation of total homocysteine, total cysteine, and methionine in normal serum and urine using capillary gas chromatography-mass spectrometry. *Anal. Biochem.* **162,** 185–196.

34. Sass, J. O. and Endres, W. (1997) Quantitation of total homocysteine in human plasma by derivatization to its N(O, S)-propoxycarbonyl propyl ester and gas chromatography-mass spectrometry. *J. Chromatog. A* **776,** 342–347.

35. Yu, S., Sugahara, K., Zhang, J., Ageta, T., Kodama, H., Fontana, M., and Dupre, S. (1997) Simultaneous determination of urinary cystathionine, lanthionine, S-(2-aminoethyl)-L-cysteine and their cyclic compounds using liquid chromatography-mass spectrometry with atmospheric pressure chemical ionization. *J. Chromatog. B* **698,** 301–307.

36. Lin, B. L., Colon, L. A., and Zare, R. N. (1994) Dual electrochemical detection of cysteine and cystine in capillary zone electrophoresis. *J. Chromatog. A* **680,** 263–270.

37. Zhou, J., O'Shea, T. J., and Lunte, S. M. (1994) Simultaneous detection of thiols and disulfides by capillary electrophoresis-electrochemical detection using a mixed-valence ruthenium cyanide-modified microelectrode. *J. Chromatog. A* **680,** 271–277.

38. Vesphalec, R., Corstjens, H., Billiet, H. A. H., Frank, J., and Luyben, K. Ch. A. M. (1995) Enantiomeric separation of sulfur- and selenium-containing amino acids by capillary electrophoresis using vancomycin as a chiral selector. *Anal. Chem.* **67,** 3223–3228.

39. Jacoby, W. B. and Griffith, O. W. (eds.) (1987) Sulfur and sulfur amino acids, in *Methods in Enzymology*, vol. 143, Academic, London and New York.

40. Walker, V. and Mills, G. A. (1995) Quantitative methods for amino acid analysis in biological fluids. *Ann. Clin. Biochem.* **32,** 28–57.

41. Ueland, P. M., Refsum, H., Stabler, S. P., Malinow, M. R., Andersson, A., and Allen, R. H. (1993) Total homocysteine in plasma or serum: methods and clinical applications. *Clin. Chem.* **39,** 1764–1779.

42. Shimada, K. and Mitamura, K. (1994) Derivatization of thiol-containing compounds. *J. Chromatog. B* **659,** 227–241.

43. Zahn, H. and Gattner, H. G. (1997) Hair sulfur amino acid analysis. *EXS* **78,** 239–258.

44. Kataoka, H., Tanaka, H., Fujimoto, A., Noguchi, I., and Makita, M. (1994) Determination of sulphur amino acids by gas chromatography with flame photometric detection. *Biomed. Chromatogr.* **8,** 119–124.

45. Kataoka, H., Takagi, K., and Makita, M. (1995) Determination of total plasma homocysteine and related aminothiols by gas chromatography with flame photometric detection. *J. Chromatog. B* **664,** 421–425.

46. Kataoka, H., Takagi, K., and Makita, M. (1995) Determination of glutathione and related aminothiols by gas chromatography with flame photometric detection. *Biomed. Chromatogr.* **9,** 85–89.

47. Takagi, K., Kataoka, H., and Makita, M. (1996) Determination of glutathione and related aminothiols in mouse tissues by gas chromatography with flame photometric detection. *Biosci. Biotech. Biochem.* **60,** 729–731.

48. Kataoka, H., Yamamoto, S., and Makita, M. (1984) Quantitative gas-liquid chromatography of taurine. *J. Chromatog.* **306,** 61–68.

49. Kataoka, H., Ohnishi, N., and Makita, M. (1985) Electron-capture gas chromatography of taurine as its N-pentafluorobenzoyl di-n-butylamide derivative. *J. Chromatog.* **339,** 370–374.

50. Kataoka, H., Ohishi, K., and Makita, M. (1986) Gas chromatographic determination of cysteic acid. *J. Chromatog.* **354,** 482–485.

51. Kataoka, H., Yamamoto, H., Sumida, Y., Hashimoto, T., and Makita, M. (1986) Gas chromatographic determination of hypotaurine. *J. Chromatog.* **382,** 242–246.

52. Okazaki, T., Kataoka, H., Fujimoto, A., Kono, K., and Makita, M. (1989) Determination of taurine in biological samples by GC with flame photometric detection. *Bunseki Kagaku* (Japanese) **38,** 401–403.

53. Makita, M., Yamamoto, S., and Kono, M. (1976) Gas-liquid chromato- graphic analysis of protein amino acids as N-isobutoxycarbonylamino acid methyl esters. *J. Chromatog.* **120,** 129–140.

54. Yamamoto, S., Kiyama, S., Watanabe, Y., and Makita, M. (1982) Practical gas-liquid chromatographic method for the determination of amino acids in human serum. *J. Chromatog.* **233,** 39–50.

55. Mills, B. J., Richie, J. P., Jr., and Lang, C. A. (1990) Sample processing alters glutathione and cysteine values in blood. *Anal. Biochem.* **184,** 263–267.

56. Kataoka, H., Imamura, Y., Tanaka, H., and Makita, M. (1993) Determination of cysteamine and cystamine by gas chromatography with flame photometric detection. *J. Pharm. Biomed. Anal.* **10,** 963–969.

16

Capillary Electrophoretic Determination of 4-Hydroxyproline

Qingyi Chu and Michael Zeece

1. Introduction

Capillary electrophoresis (CE) represents a relatively new separation technology that has gained acceptance in a wide variety of applications. A discussion of the basic theory of CE separations is beyond the scope of this presentation, but is well addressed in a number of recent texts (1–5). In principle, differential migration of analytes in a potential field is achieved as a result of individual differences in mass/charge ratios. In CE, substantial advantage in separation speed, efficiency, and resolution is derived from the very-high field strengths (e.g., 150–300 V/cm) typically used. In addition, high field strengths induce the effect of electroosmosis in the column. The walls of the silica column have a negative charge and attract hydrated counter ions from the buffer. When power is applied to the system, the positively charged ions with their associated water molecules, migrate toward the cathode with substantial velocity. This results in the flow of water termed electroosmosis. The flow proceeds from the anode to the cathode and serves as a pump. Because the electroosmotic water flow is much greater than the velocity of the analytes, all components are swept to the cathode.

Most columns used in CE have an internal diameter of 50–100 μm and can be 1 m long. High field potentials can be used because the column has a large surface area to internal volume ratio that facilitates rapid dissipation of heat. Extremely small volume samples are required for CE analysis. For example, a typical amino acid separation, such as those presented here, employ 10–20 nL sample injections. The small volume of sample injection however, also presents a challenge for detection. In order to achieve good sensitivities with UV detection, it is necessary to label the amino acids with phenylisothiocynate

From: *Methods in Molecular Biology*, vol. 159: *Amino Acid Analysis Protocols*
Edited by: C. Cooper, N. Packer, and K. Williams © Humana Press Inc., Totowa, NJ

(PITC) prior to analysis. This reaction converts the amino acids to their phenylthiohydantoin (PTH) derivatives.

The mode of CE separation used for the amino acid analysis described here employs the anionic detergent sodium dodecyl sulfate (SDS). CE performed in the presence of SDS is often termed a micellar electrokinetic chromatography (MEKC) separation *(6)*. When SDS concentrations are above 8–10 m*M*, the molecules form micelles in which the hydrophobic tails are oriented toward the center and sulfate groups are on the surface. The micelles have a strong negative charge at the pHs employed in most MEKC separations and migrate toward the anode. Their velocity, however, is less than that of the electroosmotic flow and all components are carried to the cathode. The SDS micelle thus becomes a pseudostationary phase with which analytes can interact. The resulting partitioning provides additional selectivity and the technique has been extensively used for the separation and determination of amino acids and other compounds *(7)*.

2. Materials

2.1. Equipment

1. CE units are available from several well-known vendors and the method described here was performed with an ISCO (Lincoln, NE) model 3850 capillary electropherograph. Samples are loaded into the capillary by vacuum injection and detection of separated analytes is performed at 254 nm.
2. A CAESAR (Version 4.01, 1994, Roman Scientific) from Scientific Resources (Eatontown, NJ) was used to record and analyze data.
3. A Centri-Vap (Fisher Scientific, Pittsburgh, PA) or similar device for drying small volumes in microfuge tubes.
4. An amine coated capillary (50-µm id Polymicro Technology, Phoenix AZ).

2.2. Reagents

Highest purity reagents were used throughout the experiments.

1. Phenyl isothiocyanate was obtained from Pierce (Rockford, IL).
2. Sodium hydroxide.
3. Hydrochloric acid.
4. Triethylamine.
5. Diethanolamine.
6. Malonic acid .
7. Chloro(chloro.methyl)dimethylsilane.
8. Phthalic dicarboxaldehyde.
9. 3- and 4-Hyp.
10. Cis and trans 4-Hyp.
11. Methanol.
12. Sodium phosphate (monobasic).

13. SDS.
14. Thymine (as internal standard).
15. Water used in these experiments was obtained by reverse osmosis (18 mOhm).

2.3. Amine Capillary Preparation

An amine coated capillary was used in these separations (50-μm id Polymicro Technology, Phoenix, AZ). The length to the detector was 40 cm and total column length was 70 cm. This column can be purchased from commercial vendors or can be made in the lab (*see* **Note 1**). The procedure for coating the capillary with diethanolamine was originally described by Kuhn and Hoffstetter-Kuhn *(8)* and is summarized as follows.

1. Using a 1-mL syringe fitted with an adapter, fill 1 m of fused silica capillary with 1 *M* NaOH and let stand at room temperature (20°C) for 1 h.
2. Rinse the capillary twice with water.
3. Rinse the capillary twice with methanol followed by two rinses with acetone
4. Fill the capillary with 10% (w/v) chloro(chloromethyl)-dimethylsilane in methanol and hold for 30 min at room temperature.
5. Cover the ends of the silane-filled capillary with tape and incubate at 100°C for 3 h.
6. Flush the column with first with methanol, then with water.
7. Fill the capillary with 3.0 *M* diethanolamine HCl dissolved in methanol, and incubate overnight at room temperature.

3. Methods
3.1. PTH Labeling of Hyp Standards

Standards are necessary for identification of the Hyp peak in the separation and for development of standard curves. The preparation of these standards as PTH derivatives of Hyp isomers was performed essentially as described by Chu et al. *(9)* and summarized below.

1. Dissolve 6.6 mg of Hyp in 1.0 mL of coupling buffer. This solution contains methanol-triethylamine-water in 7:1:1 (v/v) ratio. A 1.5 or 2.0 mL microfuge tube works well for holding this mixture.
2. Flush the tube with nitrogen gas and add 10 μL of PITC per 100 μL of sample.
3. Incubate the reaction for 10 min at room temperature and then dry in a Centri-Vap (Fisher Scientific).
4. The samples are stored dry at –70°C and are dissolved in 100 μL acetonitrile just prior to electrophoresis (*see* **Note 2**).

3.2. Preparation of Muscle Samples

Hyp determination can be made from isolated collagen fractions or whole muscle tissues samples. The procedures for preparing these materials for labeling is summarized as follows (*see* **Note 3**).

3.2.1. Isolated Collagen Fractions

Perimysial or other types of collagen can be isolated from skeletal muscle tissue using the well-established procedures described by Light and Champion *(10)*. Following isolation, the collagen samples are prepared as follows (*see* **Note 4**).

1. Desiccate the material by washing three times in cold (4°C) acetone. For each wash, the sample is suspended in 4 vol of cold acetone, stirred by hand and then centrifuged at 2000*g* for 10 min. The supernatant is discarded each time. Be sure to use polyethylene type tubes, because clear plastics such as polycarbonate, are attacked by the solvent.
2. Dry the sample in a hood under a stream of nitrogen.
3. Add 50 mg of collagen to a suitable hydrolysis vial.
4. Add 5.0 mL of 6 *M* HCl, draw a vacuum, and heat for 24 h at 110°C.
5. Centrifuge or filter the sample after hydrolysis to remove any insolubles.
6. Withdraw 100-µL aliquots (equivalent to approx 1.0 mg collagen sample) and place in a microfuge tube.
7. Dry the samples in a Centri-Vap. The dried samples are now ready for labeling, but can also be stored at −70°C.

3.2.2. Whole Muscle

To analyze whole muscle tissue for Hyp content, the tissue must first be dried and then hydrolyzed. The preparation of muscle samples is as follows (*see* **Note 4**).

1. Grind or finely dice 10 g of muscle.
2. Add 100–200 mg of ground muscle to a mortar and pestle prechilled with liquid nitrogen. Add liquid nitrogen slowly and grind to a powder. Keep the sample frozen hard during grinding with subsequent additions of liquid nitrogen.
3. Lyophilize the sample.
4. Add 65 mg of dried muscle powder in a suitable hydrolysis vial.
5. Add 5 mL of 6 *M* HCl, draw a vacuum and heat for 24 h at 110°C.
6. Centrifuge or filter the sample after hydrolysis to remove any insolubles.
7. Withdraw 150-µL aliquots (equivalent to about 2.0 mg muscle) and place in a microfuge tube.
8. Dry the samples in a Centri-Vap. The dried samples are now ready for labeling, but can also be stored at −70°C.

3.3. Labeling of Amino Acids from Hydrolyzed Samples

The strategy for labeling amino acids from hydrolyzed samples is based on a modification of the procedure described by Yaegaki et al. *(11)*. Amino acids from hydrolyzed tissues are labeled using a two-step procedure that involves initial reaction with *o*-phthalaldehyde (OPA) to react with primary amines followed by their precipitation under acidic conditions. In the second step, amino acids (Pro and Hyp) are reacted with PITC. The procedure is summarized as follows (*see* **Note 5**).

1. Add 100 µL of 0.5 M Na$_2$ CO$_3$, NaHCO$_3$ pH 9.5, to the dried aliquot of hydrolyzed sample. The sample should contain approx 1.0 mg of hydrolyzed material.
2. Add 100 µL (2 mg) of 20 mg/mL OPA reagent, mix, and let stand at room temperature for 15 min.
3. Precipitate the reaction products by adding of 900 µL of 1.0 M NaH$_2$PO$_4$ pH 2.5. Be sure to mix the sample well (e.g., in a vortex mixer).
4. Centrifuge at 10,000g for 5 min and the filter the supernatant through 0.45-µm spin-filters (Millipore, Bedford, MA)
5. Dry 100 µL of filtrate in the Centri-Vap.
6. Add 150 µL coupling buffer (described in **Subheading 3.1.**).
7. Flush the vial with nitrogen and add 10 µL of PITC.
8. Incubate the mixture at room temperature for 10 min and Centri-Vap to dryness.
9. The dried samples are ready for CE analysis or can be stored at –70°C.

3.4. CE Analysis

Samples are ready for analysis after the appropriate preparation steps have been performed. The procedure for CE for identification and determination of Hyp in these samples is described below (*see* **Note 6**).

1. Add 100 µL 50% acetonitrile to the tube containing the dried and labeled amino acid sample (prepared in **Subheading 3.3.**). Mix well by vortexing.
2. Centrifuge at 10,000g for 2 min.
3. Withdraw 30 µL of the sample supernatant and add 10 µL 2.0 mM thymine (internal standard).
4. Equilibrate the column filling it with the run buffer (50 mM Na-malonate, 75 mM SDS, pH 5.0).
5. The sample is then loaded to the CE using either vacuum or pressure injection following the CE manufacturer's instructions. Generally, loads will be in the order of 5 to 50 nL. The equipment manufacturer should provide information regarding determination of load volumes.
6. Set the instrument voltage at 15 kV and start the run.
7. Record the data for all peaks.
8. Identify the thymine and Hyp peaks in the separation from the procedure described below.

3.5. Quantitation of Hyp in Samples

In order to determine the amount of Hyp accurately in unknown samples, it is first necessary to identify the peak in the mixture using coinjection and then use an internal standard method for its quantitation. The procedures described here were developed for determination of Hyp in muscle tissue samples *(9)*. It

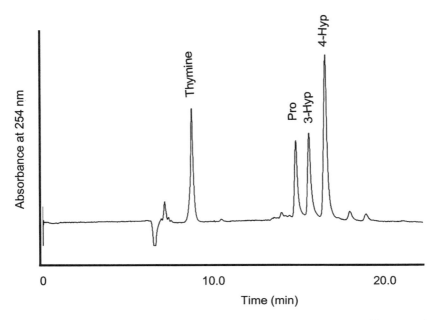

Fig. 1. Electropherogram of PTH-Proline, 3- and 4-Hydroxyproline standards. Amino acids were labeled with PITC as described in **Subheading 3.1.** to obtain the PTH derivatives. Approximately 20 nL of a mixture containing 2.0 mM each of PTH-Proline, 3- and 4-Hydroxyproline was separated by capillary electrophoresis in 50 mM Na-malonate, 75 mM SDS, pH 5.0 at 15 kV. The capillary was amine coated, 50 μm in id and 70 cm in length (40 cm to the detector). The migration of the components are identified in the figure.

is important for accurate quantitation that the Hyp peak be reasonably separated from other peaks in the sample. Thus it may be necessary to adjust parameters such as voltage or buffer composition when analyzing samples prepared from other tissues.

3.5.1. Identification of Hyp in the Separation (Electropherogram)

In order to identify the Hyp peak in a sample, the migration time must first be determined using standards (prepared in **Subheading 3.1.**) (*see* **Note 7**). A sample electropherogram illustrating the separation of proline, 3-hydroxyproline and 4-hydroxyproline is shown in **Fig. 1**. The procedure is as follows.

1. Dilute the appropriate PTH-Hyp isomer to 2.0 mM in acetonitrile.
2. Add 10 μL of 2.0 mM thymine per 30 μL of Hyp solution.
3. Mix and perform CE separation as described in **Subheading 3.4.** The conditions for separation of the standard (s) must be the same as those used for samples.

3.5.2. Coinjection of Hyp

1. Obtain the electropherogram for the unknown samples by performing the separation as described in **Subheading 3.4.**
2. Add 10 μL 2.0 m*M* Hyp to the sample and run the separation again.
3. Compare the two electropherograms. The peak with substantially increased area and height corresponds to the Hyp peak.

3.5.3. Standard Curve

1. Prepare a set of at least 6 Hyp dilutions covering the range of 0.05 to 2.0 m*M*.
2. Add 10 μL 2.0 m*M* thymine to 30 μL of each standard concentration and mix.
3. Perform the separations as described in **Subheading 3.4.** using a constant injection volume (e.g., 10 nL).
4. Determine the **Hyp peak area ratio** for each standard concentration by dividing the peak area for Hyp by that of the thymine peak.
5. Plot the **Hyp peak area ratio** vs the Hyp concentration for each point. Use a linear regression analysis of the data points. The slope of this line (Hyp concentration/peak area ratio) represents the **Conversion Factor**. The concentration of Hyp in the unknown sample is calculated from the equation below.

$$\text{Hyp peak area ratio in unk} \times \text{Conversion Factor}$$
$$(\text{Hyp conc/peak area}) = \text{Hyp Conc}$$

6. To determine the amount (moles) of Hyp in the unknown sample, the total volume of the hydrolyzed aliquot must be used. For example, assuming that an aliquot corresponding to 1.0 mg of tissue was hydrolyzed, labeled, and finally dissolved in a final volume of 100 μL then the total amount of Hyp in that sample is calculated as follows:

$$\text{Hyp Conc} \times \text{injection volume (nL)} \times \text{dilution factor}$$
$$(100 \ \mu\text{L}/10 \ \text{nL}) = \text{moles Hyp}$$

Because the amount of tissue hydrolyzed is known, the Hyp content can also be expressed as moles of Hyp per mg tissue.

4. Notes

1. Amine capillary preparation: Since the time of our original investigations, the commercial availability of amine coated capillaries has significantly increased. The quality and consistency of prepared capillaries is well worth the cost. Thus, unless the investigator is very familiar with this technology, it is recommended that amine coated capillary be obtained through a commercial vendor.
2. PTH Labeling of Hyp Standards: The labeling reaction of Hyp isomers or Pro with PITC is rather easy to do. Precautions include flushing the reaction mixture with nitrogen and protecting it from light. UV light accelerates the degradation of PTH derivatives. Storage should be in a light protected vial at –20° to –70°C. The shelf life of PTH-Hyp was only 2–3 wk. Additional peaks representing breakdown products were observed in CE separations with increasing storage.

3. This method describes procedures for application of CE to the determination of Hyp isomers and Pro in muscle tissue. It is also be applicable to the analysis of these amino acids in other tissues. Additional considerations that might be helpful in performing Hyp analysis are addressed below.

4. Whole muscle samples or isolated collagen fractions can be analyzed by this procedure. The protocol has a number of steps, but is faster and requires less sample than other (HPLC) methods. The sensitivity of this method was found to be sufficient such that determinations could be made in as little as 1 mg of sample (*9*).

5. A critical step for determination of Hyp in muscle samples of involves labeling of amino acids in hydrolyzed sample (**Subheading 3.3.**). This a two-step procedure in which amino acids (all except Hyp and Pro) are reacted with OPA. The OPA amino acids are then precipitated with 1 *M* 1.0 *M* NaH$_2$PO$_4$, pH 2.5. The insoluble amino acid derivatives are then removed by centrifugation the supernatant containing imino acids (Hyp and Pro) is subsequently reacted in the second step with PITC. However, precipitation of OPA derivatized amino acids may not be complete with NaH$_2$PO$_4$ in some samples. Alternatively, we have found that substitution of OPA with Naphthalenedicarboxaldehyde results in more efficient precipitation with NaH$_2$PO$_4$ and greater solubility for Hyp and Pro.

6. The CE separation and determination of Hyp and Pro in biological samples is relatively straightforward and faster than many HPLC approaches because the smaller sample requirements result in faster processing time. For example, this procedure contains a number of steps in which drying of the sample is necessary. Because the CE procedure requires only 100–150 µL hydrolysis aliquots, the time required to process the sample in substantially reduced. We have also successfully performed Hyp analysis with 1–5 mg samples using microwave hydrolysis containing 200–500 µL of HCl, to further reduce the processing time.

7. A critical factor in the quantitation of Hyp in biological samples by CE is the requirement for an internal standard. CE uses very small sample volumes (nL) and thus slight differences in the injection volume can result in large variations for Hyp peak areas. The use of internal standard peak area ratio method eliminates the errors caused by injection differences. The thymine peak also provides a reference point (migration time) for assessing the consistency of electropherograms.

References

1. Grossman, P. D. (1992) Factors affecting the performance of capillary electrophoresis separations: joule heating, electrosmoosis, and zone dispersion, in *Capillary Electrophoresis: Theory and Practice* (Grossman, P. J. and Colburn, J. C., eds.), Academic, London and New York, pp. 3–44.

2. Kuhn, R. and Hoffstetter-Kuhn, S. (1993) *Capillary Electrophoresis: Principles and Practice.* Springer Laboratory, New York.

3. Karger, B. L. and Foret, F. (1993) Capillary electrophoresis: introduction and assessment, in *Capillary Electrophoresis Technology* (Guzman, N., ed.), Marcel Dekker, New York, pp. 3–64.

4. Khur, W. (1993) Separation of small organic molecules by capillary electrophoresis, in *Capillary Electrophoresis: Theory and Practice* (Camilleri, P., ed.), CRC, Boca Raton, FL, pp. 65–116.

5. Morning, S. E. (1996) Buffers, electrolytes and additives for capillary electrophoresis, in *Capillary Electrophoresis in Analytical Biotechnology* (Righetti, P., ed.), CRC, Boca Raton, FL, pp. 37–60.

6. Otsuka, K., Terabe, and S., Ando, T. (1985) Electrokinetic chromatography with micellar solutions, separation of phenylthiohydantoin-amino acids. *J. Chromatog.* **332,** 219–226.

7. Matsubara, N. and Terabe, S. (1996) Micellar electrokinetic chromatography in the analysis of amino acids and peptides, in *Capillary Electrophoresis in Analytical Biotechnology* (Righetti, P., ed.), CRC, Boca Raton, FL, pp. 155–182.

8. Kuhn, R. and Hoffstetter-Kuhn. (1993) *Capillary Coatings in Capillary Electrophoresis: Principles and Practice.* Springer Laboratory, New York, pp. 162–180..

9. Chu, Q., Evans, B. T., and Zeece, M. G. (1996) Quantitative separation of 4-hydroxyproline from skeletal muscle collagen by micellar electrokinetic capillary electrophoresis. *J. Chromatog. B* **692,** 293–301.

10. Light, N. and Champion, A. E. (1984) Characterization of muscle epimysium, perimysium, and endomysium collagens. *Biochem. J.* **219,** 1017–1026.

11. Yaegaki, K., Tonzetich, J., and Ng, A. S. K. (1996) Improved high performance liquid chromatography method for quantitation of proline and hydroxyproline in biological materials. *J. Chromatog.* **356,** 163–170.

17

Total Plasma Homocysteine Analysis by HPLC with SBD-F Precolumn Derivatization

Isabella Fermo and Rita Paroni

1. Introduction

Over the last two decades, various studies have shown that moderate and persistent hyperhomocysteinemia is implicated in the development of atherosclerosis, which is responsible for 50% of all mortality and morbidity in Western countries. Considering that the most traditional risk factors for heart disease and stroke, such as plasma lipids, cigarette smoking, hypertension, obesity, and diabetes, only account for 50% of cardiovascular disease (1,2), one can understand the reason why homocysteine (Hcy) measurement is included in the list of tests for investigating the causes of atherosclerosis and thrombosis.

One of the major problems encountered in studies on the potential atherogenic role of Hcy was the development of an accurate and simple assay, capable of screening, in a normal population, subjects having a congenital predisposition to occlusive vascular disease. Several approaches have been described in literature for measuring total plasma homocysteine (tHcy), which is defined as the sum of free and protein-bound homocysteine, homocystine, and homocysteine-cysteine mixed disulfide. These procedures involve, after a reduction step, the use of gas-chromatography-mass spectrometry (GC-MS) (3), radioenzymic assay (4), and high-performance liquid chromatography (HPLC). The latter, which is the most widely applied, may be coupled with (1) spectrophotometric detection in the visible range (amino acid analyzer) after postcolumn ninhydrin derivatization (5); (2) direct electrochemical detection (6); or (3) spectrofluorimetric detection after precolumn derivatization with different labeling agents such as monobromobimane (mBrB) (7), o-phthaldialdehyde (OPA) (8), or halogensulfonyl benzofurazans (ABD-F and SBD-F) (9,10).

From: *Methods in Molecular Biology*, vol. 159: *Amino Acid Analysis Protocols*
Edited by: C. Cooper, N. Packer, and K. Williams © Humana Press Inc., Totowa, NJ

Among the HPLC procedures for tHcy analysis in plasma, methods based on SBD-F (ammonium-7-fluorobenzo-2-oxa-1,3-diazole-4-sulphonate) reaction appear to be attractive for routine use in clinical chemistry laboratories. SBD-Hcy adduct is very stable (1 wk at 4°C) (*see* **Note 1**) *(11)* and SBD-F being a thiol-specific labeling molecule, the derivatization yields a very clean chromatogric profile in comparison to mBrB, which produces interfering fluorescence byproducts *(7)*, or to OPA, which reacts unspecifically with all the amino groups of the sample *(8)*. We have described here a procedure that involves reduction of disulfide bonds by sodium borohydride (NaBH$_4$), use of mercaptopropionyl-glycine as internal standard *(12)*, precolumn derivatization with SBD-F, and isocratic analysis by reversed-phase (RP) HPLC with fluorescence detection.

2. Materials

2.1. Equipment

1. Polypropylene micro tubes (2 mL) with screw-cap (Sarstedt, Numbrecht, Germany).
2. Microfuge model 11 (Beckman, Palo Alto, CA).
3. Block heater (Asal, Milano, Italy) or oven.
4. HPLC monopump model 116 with a solvent selector valve (Beckman, Palo Alto, CA) (*see* **Note 2**).
5. Fluorimetric detector model RF 551 (Shimadzu, Kyoto, Japan).
6. Autosampler model 507 (optional device) (Beckman).
7. Column: Ultrasphere ODS (150 × 4.6-mm id, 5 μm) (Beckman).
8. Guard-column: Lichrospher 100 RP–18 (40 × 4-mm id, 5 μm) (Merck, Darmstadt, Germany).

2.2. Reagents

1. NaBH$_4$.
2. Dimethyl sulfoxide (DMSO).
3. 7-fluorobenzo-2-oxa-1,3-diazole-4-sulphonate (SBD-F).
4. Boric acid.
5. ethylenediaminetetracetic acid (EDTA).
6. L-homocystine (Hcy-Hcy).
7. Sodium acetate.
8. Sodium hydroxide (NaOH).
9. 70% perchloric acid (PCA).
10. Methanol (CH$_3$OH).
11. Acetonitrile (CH$_3$CN).
12. Glacial acetic acid.
13. Mercaptopropionylglycine (IS).
14. Bidistilled water (H$_2$O).

2.3. Solutions

1. Reducing solution: Prepare 3 mol/L $NaBH_4$ solution by dissolving 0.285 mg of $NaBH_4$ in 2.5 mL of NaOH 0.1 mol/L, then mix with DMSO (2:1, v/v) and store in a glass vial at 4°C. Prepare this solution fresh every week.
2. Deproteinizing solution: Prepare 0.6 mol/L PCA solution by mixing 25 mL PCA (70%) with 0.186 g EDTA and diluting to 500 mL with H_2O. This solution is indefinitely stable when stored at 4°C (*see* **Note 3**).
3. Buffer solution: Prepare 0.2 mol/L boric acid solution with 4 mmol/L EDTA, pH 9.5 by dissolving 12.37 g boric acid and 1.49 g EDTA in 1 L H_2O. Warm to dissolve. Adjust to pH 9.5 with concentrated NaOH while still warm. Cool to room temperature before use and store this solution at room temperature.
4. Neutralizing solution: Prepare NaOH 2 mol/L by dissolving 80 g NaOH in 1 L H_2O.
5. Derivatizing solution: Prepare 4.2 mmol/L SBD-F solution by mixing 1 mg SBD-F with 1 mL of buffer solution. Store at 4°C in a dark vial for 1 d.
6. Internal standard solution: Mix 1 mg of IS with 1 mL HCl 0.1 mol/L and 9 mL H_2O. Store 200-µL aliquots at –20°C up to 6 mo. To use, dissolve each aliquot with 600 µL solution 3 diluted 1:1 with H_2O. Store at 4°C for 1 wk.
7. Homocystine standard solution: Prepare solution 200 µmol/L by dissolving in a 100 mL volumetric flask 5.4 mg of Hcy-Hcy in 5 mL HCl 0.1 mol/L and diluting to 100 mL with H_2O. Freeze 300-µL aliquots at –20°C for 6 mo. Use each aliquot only once and then discard.
8. Stock HPLC eluents: Prepare sodium acetate solution 0.2 mol/L by dissolving 27.22 g sodium acetate trihydrate in 1 L H_2O. Dilute 11.5 mL glacial acetic acid with 1 L H_2O and adjust to pH 4.0 with the sodium acetate solution ($\approx$3:1) to obtain 0.2 *M* sodium acetate buffer. Store all these solutions at 4°C.
9. Working HPLC eluents: Mobile phase A: add 20 mL of CH_3OH to 980 mL of buffer 8 (2% CH_3OH). Mobile phase B: add 300 mL of CH_3OH to 700 mL of solution 8 (30% CH_3OH).

3. Methods

3.1. Preanalytical Steps

Collect whole blood specimens in sterile Vacutainer® tubes containing sodium citrate (Becton Dickinson, Rutherford, NJ). Keep the samples on ice until centrifugation at 3000*g* for 10 min at 4°C, then immediately freeze the plasma specimens at –20°C until analysis (*see* **Note 4**).

3.2. Procedure (see Flow Chart 1)

1. Reduction step. Mix 100 µL plasma sample with 50 µL of internal standard solution and 20 µL of $NaBH_4$ solution (reducing solution) in a polypropylene tube with screw-cap. Incubate at 50°C for 30 min (*see* **Note 5**).

Flow Chart 1
SBD-F Precolumn Derivatization Method

• Plasma collection	⇒	Anticoagulant: sodium citrate, eparin, or EDTA
		- Blood sample in ice
		- Prompt separation of plasma
• IS addition	⇒	100 µL mercaptopropionylglycine solution
• Reduction of disulfide bonds	⇒	by 20 µL 3 mol/L NaBH4 solution (30 min at 50°C)
• Plasma deproteinization	⇒	by 100 µL 0.6 mol/L PCA in 1 mmol/L EDTA
• Plasma neutralization	⇒	by 20 µL 2 mol/L NaOH
• Derivatization	⇒	by 50 µL SBD-F solution (60 min at 60°C)
• HPLC separation	⇒	isocratic elution (14 min); reversed-phase C18 column
• Fluorimetric detection	⇒	$\lambda ex = 385$ nm; $\lambda em = 515$ nm

2. Deproteinization step. Add 100 µL 0.6 mol/L PCA (solution 2). Vortex, then centrifuge at 12000g for 10 min.
3. Derivatization step. Transfer 100 µL of supernatant in an Eppendorf tube and add 200 µL of 0.2 mol/L borate buffer pH 9.5 (buffer solution), 20 µL of 2 mol/L NaOH (neutralizing solution), and 50 µL of SBD-F solution (derivatizing solution). Vortex and then incubate the samples at 60°C for 1 h. Remove and allow vials to cool to room temperature before injecting (*see* **Note 6**).
4. Transfere derivatized samples into the autosampler vials (optional device) and inject 10 µL into the HPLC (*see* **Notes 7** and **8**).

3.3. Calibration Curve

1. For each batch of samples to be run on HPLC, dilute the 200 µmol/L homocystine stock solution to get the final concentrations: 100, 50, and 25 µmol/L (use solution 3 diluted 1:1 with H_2O).
2. Spike 100 µL of pooled plasma with 10 µL of each of these hcy-hcy standard solutions to obtain a calibration curve in the 5, 10, 20, 40 µmol/L range (Hcy molar equivalents) (*see* **Notes 7** and **9**). Plot the relative fluorescence intensities of SBF-Hcy/IS (y) as a function of Hcy concentrations (x), and perform the least square analysis to obtain the regression equation.
3. The unknown tHcy concentration in a plasma sample is determined by this calculation:

$$[\text{tHcy}]_{\text{sample}} \text{ µmol/L} = (\text{Hcy peak area : IS peak Area})_{\text{sample}}/\text{slope of calibration curve}$$

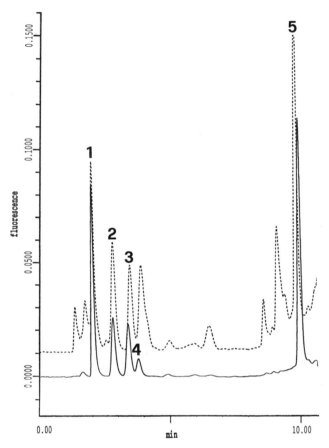

Fig. 1. Chromatograms of: (—) plasma sample with tHcy level of 23.0 µmol/L; (---) Bio-Rad Liquicheck™ Homocysteine Control-2 (lot. no. 18312, Bio-Rad Diagnostic Group, Irvine, CA). The separation was carried out on a Beckman Ultrasphere ODS column (150 × 4.6-mm id; 5 µm). (peak 1 = cysteine; peak 2 = homocysteine; peak 3 = cysteinylglycine; peak 4 = glutathione; peak 5 = mercaptopropionylglycine).

3.4. Chromatographic Analysis

1. Fluorimetric detector: λex = 385 nm; λem = 515 nm.
2. Run time: 14 min.
3. Equilibrate the column with 100% mobile phase A at 1.2 mL/min and 0.2 min after the analysis started, switch the selector valve to 100% eluent B.
4. At 4 min increase flow rate to 1.4 mL/min over 0.5 min, and at 4.5 min, switch back the valve to 100% eluent A.
5. At 12 min, decrease the flow rate to 1.2 mL/min in 0.5 min and at 14 min you are ready for the next injection.

HPLC Run Program

Time (min)	Flow (mL/min)	Duration (min)	Mobile phase
0	1.2	—	A
0.2	1.2	—	B
4.0	1.4	0.5	B
4.5	1.4	—	A
12	1.2	0.5	A
14	STOP	—	A

The chromatogram in **Fig. 1** shows the following peaks: cysteine (1), homocysteine (2), cysteinylglycine (3), glutathione (4), and mercaptopropionylglycine (5).

4. Notes

1. The fluorescence SBD-Hcy adduct, being photosensitive, has to be stored in the dark where is stable for 1 wk at 4°C *(11)*.

2. This method requires the use of a simple monopump HPLC, but with a solvent selector valve. If the solvent switching is not available, the IS will be eluted later and the analysis will last too long. As an alternative, the method can be performed without the addition of IS, and the chromatographic separation will be performed using only the mobile phase A at flow rate 1.2 mL/min. In this case, we have to be careful, above all, during the reduction step and the use of a surface active agent is recommended. In addition, we suggest running the plasma calibration curve at least three times per day in order to check the performance of own detector system.

3. If the deproteinizing solution shows an amount of precipitate on the bottom of the glass vial, we are probably using out-of-date reagents. The reductant must to be completely dissolved in the vial before its use.

4. We adopted sodium citrate as anticoagulant as it is commonly used by the Coagulation Service of our Institute; no differences in tHcy plasma levels were found after EDTA or heparin treatment *(8)*. It is important to know that differences in blood-sample handling can cause significantly different results in tHcy measurement. During whole blood storage at room temperature, Hcy concentration tends to increase over time because of the protracted production and release of Hcy by blood cells. This process is slowed down when blood samples are left on ice. There are no problems for plasma samples because tHcy in this biological fluid is stable for at least 4 d at room temperature, for several weeks at 4°C and for years at −20°C *(7)*.

5. Sodium borohydride is a potent reductant, but because this reaction involves the formation of gas and sample foaming, the use of surface active agents such as amyl alcohol is also suggested. In our method, this addition is avoided thanks to (1) the use of 2-mL conical bottom tubes with screw-caps and (2) the addition of an internal standard. Pay attention during PCA deproteinization also: as the addition of an acid sparks off sample foaming. For this reason, we prefer to utilize tubes with an internal volume of 2 mL, at least.

6. Derivatization reaction can be directly performed in the autosampler vials if these tubes are heat-resistant and suitable for the block heater (incubation equipment).
7. The lowest limit for reliable quantification (LOQ) of the method is 0.85 µmol/L, corresponding to 1.7 pmol injected *(12)*, and the analysis is carried out by utilizing a simple fluorometric detector on line with the HPLC apparatus.
8. The normal ranges for plasma tHcy differ from one laboratory to another, and these discrepancies may be related to the procedure or technique used or the criteria chosen to select the control subjects. However, tHcy plasma levels between 5.0 and 15.0 µmol/L are usually considered as normal *(13)*. By applying this procedure, we found the following reference values: 7.65 ± 2.36 and 8.9 ± 2.14 µmol/L for females ($n = 40$) and males ($n = 40$), respectively *(8)*.
9. Plasma calibration curve must be used instead of aqueous calibrators because the latter give a linear equation with a different slope (–20%), by showing highlighting the notable influence of the matrix on SBD-F thiol derivatization *(9,12)*. We can use Hcy standard instead of Hcy-Hcy, but, in this case, care must be taken with the purity of the Hcy powder *(14)*. All standards must be stored at –20°C.

References

1. Masser, P. A., Taylor, L. M. D., and Porter, J. M. (1994) Importance of elevated plasma homocysteine levels as a risk factor for atherosclerosis. *Ann. Thorac. Surg.* **58,** 1240–1246.
2. Stampfer, M. J. and Malinow, M. R. (1995) Can lowering homocysteine levels reduce cardiovascolar risk? *N. Engl. J. Med.* **332,** 328–329.
3. Sass, J. O. and Endres, W. (1997) Quantitation of total homocysteine in human plasma by derivatization to its N(O, S)-propoxycarbonyl propyl ester and chromatography-mass spectrometry analysis. *J. Cromatog. A* **776,** 342–347.
4. Frantzen, F., Faaren, A. L., Alfheim, I., and Nordhei, A. K. (1998) Enzyme conversion immunoassay for determining total homocysteine in plasma or serum. *Clin. Chem.* **44,** 311–316.
5. Andersson, A., Brattstrom, L., Isaksson, B., and Israelsson, B. (1989) Determination of homocysteine in plasma by ion-exchange chromatography. *Scand. J. Clin. Invest.* **49,** 445–449.
6. Malinow, M. R., Kang, S. S., and Taylor, L. M. (1989) Prevalence of hyperhomocyst(e)inemia in patients with peripheral arterial occlusive disease. *Circ. Res.* **79,** 1180–1188.
7. Fiskerstrand, T., Refsum, H., Kvalheim, G., and Ueland, P. M. (1993) Homocysteine and other thiols in plasma and urine:automated determination and sample stability. *Clin. Chem.* **39,** 263–271.
8. Fermo, I., Arcelloni, C., De Vecchi, E., Vigano', S., and Paroni, R. (1992) HPLC method with fluorescence detection for the determination of total homocyst(e)ine in plasma. *J. Chromatog. (Symp)* **593,** 171–176.
9. Vester, B. and Rasmussen, K. (1991) HPLC method for rapid and accurate determination of homocysteine in plasma and serum. *Eur. J. Clin. Chem. Clin. Biochem.* **29,** 549–554.

10. Feussner, A., Rolinski, B., Weiss, N., Deufel, T., and Rosher, A. (1997) Determination of total homocysteine in human plasma by isocratic HPLC. *Eur. J. Clin. Chem. Clin. Biochem.* **35,** 687–691.
11. Imai, K., Toyo'oka, T., and Watanabe, Y. (1983) A novel fluorogenic reagent for thiols:ammonium 7-fluorobenzo-2-oxa-1, 3-diazole-4-sulfonate. *Anal. Biochem.* **128,** 471–473.
12. Fermo, I., Arcelloni, C, Mazzola, G., D'Angelo, A., and Paroni, R. (1998) HPLC method for measuring total plasma homocysteine levels. *J. Chromatog. B* **719,** 31–36.
13. Ueland, P. M., Refsum, H., Stabler, S. P., Malinow, M. R., Andersson, A., and Allen, R. H. (1993) Total homocysteine in plasma or serum: methods and clinical application. *Clin. Chem.* **39,** 1764–1779.
14. Dudman, N. P. B., Guo, X. W., Crooks, R., Xie, L., and Silberberg, J. S. (1996) Assay of plasma: light sensitivity of the fluorescence SBD-F derivative, and use of appropriate calibrators. *Clin. Chem.* **42,** 2028–2032.

18

Determination of Early Glycation Products by Mass Spectrometry and Quantification of Glycation Mediated Protein Crosslinks by the Incorporation of [14C]lysine into Proteins

Malladi Prabhakaram, Beryl J. Ortwerth, and Jean B. Smith

1. Introduction

The Maillard reaction, popularly known as nonenzymatic glycosylation (NEG) or glycation, is a complex chemical reaction and occurs in vivo between reactive aldose or ketose sugars and protein-bound free amino groups *(1)*. NEG has been implicated in diabetic or age-related complications *(2)*, Alzheimer's disease *(3)*, and also in cataract formation *(4)*. In vivo, any protein with free amino groups can react with reducing sugars via the Maillard reaction. However, the extent of damage caused by NEG is amplified in diabetic tissues (because of the elevated levels of blood sugar levels) and also in proteins with a long half-life like collagen and lens crystallin proteins *(1,2)*. Therefore, determination of early glycation products and protein crosslinks produced because of glycation will be a significant aspect to investigate the extent of damage caused by NEG in vivo.

As shown in **Fig. 1**, the Maillard reaction involves the initial interaction of a reducing sugar (like glucose in **Fig. 1**) with α-/ϵ- amino groups of protein-bound amino acids, with the ϵ- amino group on lysine as the preferential site. An initial labile Schiff base formed during the reaction undergoes Amadori rearrangement and yields a stable ketoamine derivative. Following dehydration and subsequent chemical modification, some of the Amadori products accumulate with time as advanced glycation end products (AGEs). Though it has been proposed that AGEs have a significant impact in age-related or diabetic

From: *Methods in Molecular Biology*, vol. 159: *Amino Acid Analysis Protocols*
Edited by: C. Cooper, N. Packer, and K. Williams © Humana Press Inc., Totowa, NJ

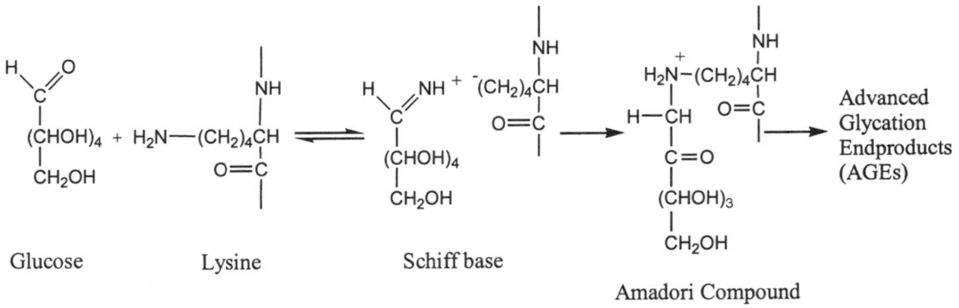

Fig. 1. Reaction scheme for the Maillard reaction (nonenzymatic glycation). Glucose is shown as the model sugar and lysine is shown as the model amino acid.

complications *(1–3)*, it is still a debate to understand the mechanism involved in the development of glucose dependent diabetic complications. This is partly because of the slow reaction property of glucose with model proteins in vitro *(5)*. Previously, we have shown that the oxidation products of ascorbic acid (which is relatively higher than glucose in human lens) can glycate and crosslink lens proteins at an accelerated rate compared to glucose under similar in vitro conditions *(6–8)*, suggesting a possible role for ascorbate mediated glycation in vivo.

Characterization and quantification of glycation crosslinks has been the subject of study for decades. Several methods have been developed in the past to determine early and advanced glycation products. These methods include borohydride reduction *(9)*, thiobarbituric acid assay *(10)*, the fructosamine assay *(11)*, periodate assay *(12)*, size-exclusion chromatography *(13)*, ion–exchange chromatography *(14)*, boronate-affinity chromatography *(15)*, and immuno detection *(16)*. With the exception of immunological methods, almost all the glycation assay methods are based on chemical methods, and are nonspecific. Additionally, in vitro assay methods do not guarantee the detection of similar compounds in vivo. For example, some of the reagents used in borohydride reduction, thiobarbituric acid assay, acid/base hydrolysis, or periodination can nonspecifically interact with nonglycated products in the assay system and generate faulty reaction products that are counted as glycated products. Other examples are the fructosamine assay and phenylboronate assay, which are based on the detection of serum albumin reacted with glucose (aldose sugar). Both these assay methods failed to detect serum albumin glycated by fructose (ketose sugar) in diabetics *(17)*. Although measurement of fluorescent compounds like pentosidine *(18)* in glycated proteins is an alternate choice, not all AGEs (e.g., carboxymethyl lysine, **ref.** *19*) are fluorescent in nature. Another drawback in estimation of fluorescent AGEs is the highly crosslinked or insoluble nature of glycated proteins in diabetic or aged tissues. Therefore, alternate assay methods to detect and quantitate the early and advanced glycation end products represent a needed and valid approach to estimate the damage caused by nonenzymatic glycation in vivo.

Recently, we have shown that mass spectrometry (MS) can be effectively used to detect and determine the early glycation products formed in a model glycation system *(20)*. MS has successfully been used to precisely determine glycated proteins or peptides in model glycation systems *(21–23)* (*see* **Note 1**). Although sodium dodecyl sulfate-polyacrylamide gel electrophoresis (SDS-PAGE) is a widely used method to observe crosslinked proteins, high molecular-weight proteins (>100 kDa) that are formed because of glycation are poorly resolved on polyacrylamide gels. These high molecular-weight proteins compress as a single band on the top of the gel and scan as a single peak by densitometric gel scanners. In order to quantitate glycation crosslinks, we have developed an in vitro assay method based on the incorporation of [^{14}C]lysine into proteins during glycation *(24)*. Because AGE inhibitors like aminoguanidine and semicarbazide inhibit the incorporation of [^{14}C]lysine into proteins *(24)*, this assay can be used to compare the crosslinking ability of various sugars, and it can also be used to test the ability of specific amino acids to act as donors for crosslinking to protein (*see* **Note 2**).

In this chapter, we describe a method to determine early glycation products by MS and also a method to quantitate the glycation crosslinks in model proteins by the incorporation of [^{14}C]lysine into proteins.

2. Materials

2.1. Equipment

1. High-performance liquid chromatography (HPLC) system + ultraviolet (UV) detector.
2. Vydac C18 HPLC column (10 × 250 mm) (Vydac, The Separations Group, Hesperia, CA).
3. Nylon filters (Gelman Scientific, Ann Arbor, MI).
4. 1.5-mL sterile amber vials (Pierce Chemicals, Rockford, IL).
5. Metal ion-free Eppendorf tubes (Perfector Scientific, Atascadero, CA).
6. FAB-MS (Varian, Netherlands).
7. ESI-MS (Varian).

2.2. Reagents

1. Diethylenetriaminepentaacetic acid (DTPA), [1-^{12}C]L-threose and other biochemicals can be obtained from Sigma Chemical Co. (St. Louis, MO). [1-^{13}C]L-threose can be obtained from Omicron Chemicals, South Bend, IN, and α-N-t-Boc-lys-ala-ala peptide can be obtained from Bachem, Inc. (Philadelphia, PA).

2. Radioactive amino acids like [^{14}C]lysine (specific activity 165 mCi/mmol) and [^{14}C]leucine (specific activity 315 mCi/mmol) can be obtained from NEN Research Products (Boston, MA) or ICN Biomedicals Inc. (Costa Mesa, CA). Ultrapure lysozyme is from AMRESCO, Solon, OH. Model sugars like glucose, ascorbic acid, dehydroascorbic acid, or L-threose and RNase can be obtained from Sigma Chemical Co. Trichloroacetic acid (TCA) can be obtained from Fisher Scientific (Pittsburgh, PA).
3. Metal-free deionized water should be used to prepare buffers and other reagents.

2.3. Solutions

1. Phosphate-DTPA buffer: 0.1 M sodium phosphate buffer, pH 7.0, containing 0.1 mM DTPA.
2. Stock A: 1 M Boc-KAA peptide in phosphate -DTPA buffer.
3. Stock B: Threose in phosphate -DTPA buffer.
4. Solvent A: 0.1% TFA.

3. Methods

3.1. Glycation of t-Boc-lys-ala-ala (Boc-KAA) Peptide by Threose [^{12}C and 1–^{13}C]

1. Prepare 0.1 M sodium phosphate buffer, pH 7.0 containing 0.1 mM DTPA as the chelator (referred as phosphate-DTPA buffer) in deionized water. Prepare stock solutions of 1 M Boc-KAA peptide and threose in phosphate -DTPA buffer just before use.
2. Prepare a reaction mixture containing peptide and threose (1:10 on a molar basis) by mixing 20 µL of 1 M peptide and 200 µL of 1 M threose in 1 mL of phosphate-DTPA buffer. Also prepare control reaction mixtures without peptide and without sugar in 1 mL of phosphate-DTPA buffer.
3. Immediately after the addition of reactants, filter the reaction mixtures through 0.2-µm sterile nylon filters (Gelman Scientific) into sterile amber vials of 1.5-mL capacity (Pierce Chemicals). Use a laminar flow hood for this purpose.
4. Incubate the vials with reaction mixtures in an incubator maintained at 37°C.
5. To observe the reaction products that accumulate with time, withdraw aliquots (equivalent to approx 100 µg peptide) at different intervals (0–72 h) (*see* **Note 3**) and freeze the aliquots immediately at –80°C.

3.2. Preparation of Samples for Mass Spectrometry

1. Thaw aliquots frozen at –80°C (**step 5, Subheading 3.1.**) at room temperature and inject on a semipreparative Vydac C18 RPHPLC column (10 × 250 mm), which has been preequilibrated with 0.1%TFA in water (solvent A) (*see* **Note 4**). Set the absorbance on UV detector at 210 nm to monitor the absorbance of the peptide and reaction products.
2. Elute reaction products that are not bound to HPLC column in 5 column volumes of solvent A. This is followed by the elution of reaction products that are bound

to HPLC column in approx 10 column volumes of solvent B (60% acetonitrile and 0.1% TFA in water).
3. For accurate mass determination, collect the reaction products in metal-free (*see* **Note 5**) Eppendorf tubes (1.5-mL/tube/min) and combine the fractions with maximum absorbance at 210 nm (*see* **Note 6**).

3.3. FABMS Analysis of Modified Peptides

1. Dissolve the dried samples (**step 3**, **Subheading 3.2.**) in 5–10 μL of methanol-water and add 90% formic acid (Spectrapure) to a final concentration of 22% (*see* **Note 7**). Remove a 1.0-μL aliquot and mix immediately with a drop of a 1-thioglycerol and place on the MS instrument probe (*see* **Note 8**).
2. Insert the probe into a mass spectrometer and record spectra over a m/z range of 300–2000 at a resolution of 2000 (*see* **Notes 9** and **10**).

3.4. Electrospray Ionization Mass Spectrometry (ESIMS) of Boc-KAA Peptide Modified During Glycation

1. Dissolve the samples obtained from HPLC (**step 3**, **Subheading 3.2.**) in a solution of 50:50 acetonitrile:water with 0.1% TFA (*see* **Note 7**) and deliver to the analyzer at a flow rate of 5 μL/min in the same solution (*see* **Note 11**).
2. Record the spectra over a m/z range 300–2000 and analyze the data with massLinx software or any other appropriate software.

3.5. Quantification of Glycation Crosslinks in Proteins by the Incorporation of [14C]Lysine into Proteins During Glycation

We describe below an assay method to determine the glycation-dependent crosslinks formed between [14C]lysine and lysozyme in the presence of a model sugar, L-threose (*see* **Note 12**). To measure the reactivity of the α-amino group alone with lysozyme/RNase during glycation, [14C]leucine can be used instead of [14C]lysine in the reaction mixture. Adequate care should be taken to handle and dispose of any radioactive waste according to NRC guidelines.

1. Prepare 0.1 *M* sodium phosphate buffer, pH 7.0, containing 0.1 m*M* DTPA as the chelator (referred as phosphate-DTPA buffer) in deionized water. Just before use, prepare stock solutions of lysozyme or RNase (25 mg/mL) and L-threose or reactive sugar of choice (1 M) in phosphate-DTPA buffer.
2. Dilute the stock solution of [14C]lysine to a working stock of 50 μCi/100 μL in deionized water and keep on ice.
3. A typical 1-mL reaction mixture is prepared as described below (*see* **Note 13**).

Lysozyme/RNase (25 mg/mL)	[14C]lysine 50 μCi/100 μL	Phosphate-DTPA buffer	L-threose (1 *M*)
200 μL	10 μL	770 μL	20 μL

4. Immediately after the addition of L-threose, sterile filter each reaction mixture through a 0.2-μm nylon filter into an amber vial (1.5-mL capacity) and incubate in an incubator maintained at 37°C.

5. Withdraw aliquots in duplicate at different intervals from 0–72 h and immediately freeze the aliquots at –80°C until use.

3.6. Measurement of Radioactivity
in the Aliquots Withdrawn from Reaction Mixtures

Because free lysine readily adheres to proteins and methods like dialysis or size-exclusion chromatography do not completely remove lysine nonspecifically bound to proteins, we routinely used the filter disk method of Mans and Novelli *(26)* to determine the protein-bound radioactivity. According to this method, proteins will be precipitated by cold TCA on a 3 MM Whatman filter paper disk and any unbound or low molecular-weight compounds will be washed off prior to determine the protein-bound radioactivity. The proteins bind to the fiber of the filter disk and remain on the filter paper during the washing procedure with TCA.

1. Aliquots frozen at –80°C should be thawed prior to use. Spot 10 mL of each aliquot (in triplicate) on 3 MM Whatman filter paper disks and air dry the filter paper disks under a fume hood (filter paper disks can conveniently suspended with pins on a Styrofoam box lid to air dry).
2. After the reagents have been absorbed into the disks, drop the filter paper disks with pins still inserted into a solution of ice-cold 10% TCA while stirring with a magnetic stirrer and wait 20 min (*see* **Note 14**). A perforated 250-mL Plastic beaker (a cork borer can be used to make holes in the plastic beaker) can be used to hold the filter paper disks and place the plastic beaker in a 500-mL glass beaker containing 10% TCA.
3. Transfer the plastic beaker with filter paper disks to a glass beaker containing 5% TCA (room temperature) and stir for 20 min. Following this, transfer the plastic beaker with filter paper disks to a glass beaker containing 5% TCA maintained at about 75°C and stir for 15 min. Hot TCA assures the measurement of incorporation into protein only.
4. Transfer the plastic beaker with filter paper disks to a mixture of ethanol-anhydrous ether (2:1) under stirring and wait for 15 min. This step is required to remove residual TCA from the filter paper disks.
5. Transfer the filter paper disks carefully to a plastic weighing boat or to a sheet of aluminium foil and air-dry the filter paper disks in the fume hood or under a heat lamp.
6. Remove the pins, transfer the dried filter paper disks to a vial containing radioactive scintillation cocktail (e.g., Ultima gold, Packard, Meriden, CT) and count radioactivity in a liquid scintillation counter. Calculate the amount of radioactivity present on each disk and determine the amount incorporated into lysozyme or RNase at different intervals during glycation reaction.

4. Notes

1. Determination of molecular weights of glycated peptides by either fast atom bombardment mass spectrometry (FABMS) or by electrospray ionization mass spec-

trometry (ESIMS) is an effective method of establishing whether a peptide has become glycated, how many sugar moieties are attached to the peptide, and whether there is evidence of crosslinks. Typical mass accuracy for determinations by both FABMS and ESIMS is about ± 0.2 Daltons. Although both techniques can be used, each has its advantages and disadvantages. FABMS is customarily used for peptides with molecular masses of 250–4000. The mass range for ESIMS is much larger. ESIMS can be used to determine molecular masses of both peptides and proteins.

2. Although [^{14}C]lysine can be used to measure glycation dependent crosslinking of [^{14}C]lysine to model proteins, N-α-acetyl[^{14}C]lysine can be used to estimate the glycation crosslinks formed because of the specific interaction of ε-amino group of lysine with proteins during glycation *(27)*. To measure specific lysine-amino acid crosslinks, homopolymers of amino acids like polylysine, polyarginine or polyhistidine can also be used.

3. L-threose is a highly reactive tetrose sugar and its reaction with free amino groups is generally completed in only a few days *(25)*. Therefore, L-threose is selected as a model sugar in the experiments. If threose is substituted with sugars like glucose or fructose, aliquots should be withdrawn at weekly intervals or until the reaction products reach to a plateau.

4. FABMS or ESIMS require that the sample should be free of nonvolatile salts containing cations such as sodium and potassium. These cations can replace the protons on some peptides, leading to many additional charged states. For this reason, it is recommended that the last step in the isolation of the peptides be RP-HPLC. Dialysis usually does not adequately remove interfering salts. Because of the extreme sensitivity to sodium ions, samples should be collected and stored in plastic tubes after the final RP desalting.

5. Because sodium ion impurities in samples injected on MS can lead to inaccurate mass spectra, care should be taken to select metal ion-free Eppendorf tubes. We routinely used plastic tubes obtained from Perfector Scientific, Atascadero, CA.

6. The most important information for someone submitting samples for mass spectrometric analysis is how much sample is required for analysis, and that the samples should not contain too many components and be free of salts.

7. For both FABMS and ESIMS, the sample is usually dissolved in an acidic solvent, such as 0.1% TFA or acetic acid before introduction into the mass spectrometer.

8. FABMS analysis is performed by mixing the dissolved sample with a matrix such as glycerol/thioglycerol or nitrobenzyl alcohol, and placing this mixture on the probe that is inserted into the instrument. The peaks caused by the matrix contribute to higher background signals, making this technique less sensitive than ESIMS, particularly at masses below 300 Daltons.

9. We have used a range between 300–2000, because the peptide we used in this study has a molecular mass of 389. An appropriate range should be selected based on the size and mass of protein or peptide in use.

10. One advantage of FABMS is that, in the positive ion mode, a monoprotonated peptide is obtained, so the spectrum shows an MH$^+$ ion, which is the molecular

mass of the peptide plus one Dalton. The molecular mass of the peptide is readily apparent. In contrast, ionization of the sample in ESIMS produces several charged states of each peptide, resulting in a spectrum showing peaks with the mass/charge ratio of each of the charged states.

11. For ESIMS, the dissolved sample is introduced into the mass spectrometer as a continuous flow of solution. Although the amount of sample required depends very much on the sample characteristics, about one nanomole of peptide is typical for a FABMS analysis, whereas 100 picomole is usually sufficient for ESIMS. Detection limits 1000-fold lower may be achieved in special cases.

12. One can also use radioactive-labeled proteins instead of [^{14}C]lysine to measure the incorporation of labeled sugars into proteins.

13. Always set up a reaction without sugar to correct for any nonspecific interaction of [^{14}C]lysine with proteins. Also minimize the volume (50–100 µL) of samples with [^{14}C]lysine to apply on the filter paper disks.

14. To show that no carryover of [^{14}C]lysine from one disk to others in between washings of the disks, include plain filter paper disks during the TCA wash procedure.

References

1. Monnier, V. M. (1989) *Toward a Maillard Reaction Theory of Aging* (Baynes, J. W. and Monnier, V. M., eds.). Alan R. Liss, New York.

2. Baynes, J. W. (1996) The role of oxidation in the Maillard reaction in vivo, in *The Maillard Reaction: Consequences for the Chemical and Life Sciences* (Ikan, R., ed.), Wiley, New York.

3. Markesbery, W. R. (1997) Oxidative stress hypothesis in Alzheimer's disease. *Free Rad. Biol. Med.* **23,** 134–147.

4. Monnier, V. M. and Cerami, A. (1983) Detection of nonenzymatic browning products in the human lens. *Biochim. Biophys. Acta* **760,** 97–103.

5. Ortwerth, B. J., Speaker, J. A., Prabhakaram, M., Lopez, M., Li, E., and Feather, M. S. (1994) Ascorbic acid glycation: the reactions of L-threose in lens tissue. *Exp. Eye Res.* **58,** 665–674.

6. Prabhakaram, M. and Ortwerth, B. J. (1991) The Glycation-associated crosslinking of lens proteins by ascorbic acid is not mediated by oxygen free radicals. *Exp. Eye Res.* **53,** 261–268.

7. Prabhakaram, M. and Ortwerth, B. J. (1992) The glycation and cross-linking of isolated lens crystallins by ascorbic acid. *Exp. Eye Res.* **55,** 451–459.

8. Prabhakaram, M. and Ortwerth, B. J. (1992) Glycation of MP26 and MP22 in bovine lens membranes. *Biochem. Biophys. Res. Comm.* **185,** 496–504.

9. Kennedy, D. M., Skillen, A. W., and Self, C. H. (1993) Colorimetric assay of glycoprotein glycation free of interference from glycosylation residues. *Clin. Chem.* **39,** 2309–2311.

10. Parker, K. M., England, J. D., Da Costa, J., Hess, R. L., and Goldstein, D. E. (1981) Improved colorimetric assay for glycosylated hemoglobin. *Clin. Chem.* **27,** 669–672.

11. Johnson, R., Metcalf, P. A., and Baker, J. R. (1982) Fructosamine: A new approach to the estimation of serum glycosylprotein. An index of diabetic control. *Clin. Chim Acta* **127,** 87–95.

12. Ahmed, N. and Furth, A. J. (1991) A microassay for protein glycation based on the periodate method. *Anal. Biochem.* **192,** 109–111.
13. Yatscoff, R. W., Tevaarwerk, G. J. M., and McDonald, J. C. (1984) Quantification of nonenzymically glycated albumin and total serum protein by affinity chromatography. *Clin. Chem.* **30,** 446–449.
14. Castagnola, M., Caradonna, P., Bertollini, A., Cassiano, L., Rossetti, D. V., and Salvi, M. L. (1985) Determination of the non-enzymatic glycation of hemoglobin by isoelectrofocusing of its globin chains. *Clin. Biochem.* **18,** 327–331.
15. Middle, F. A., Bannister, A., Bellingham, A. J., and Dean, P. D. G. (1983) Separation of glycosylated haemoglobins using immobilized phenylboronic acid. *Biochem. J.* **209,** 771–779.
16. Ikeda, K., Higashi, T., Sano, H., Jinnouchi, Y., Yoshida, M. Araki, T., Ueda, S., and Horiuchi, S. (1996) N-ε-(Carboxymethyl) lysine protein adduct is a major immunologicalepitope in protein modified with advanced glycation end products of Maillard reaction. *Biochemistry* **35,** 8075–8083.
17. Ahmed, N. J. and Furth, A. J. (1992) Failure of common glycation assays to detect glycation by fructose. *Clin. Chem.* **38,** 1301–1303.
18. Sell, D. R., and Monnier, V. M. (1989) Structure elucidation of a senescence crosslink from human extracellular matrix. *J. Biol. Chem.* **264,** 21597–21602.
19. Ahmed, M. U., Thorpe, S. R., and Baynes, J. W. (1989) Identification of N-epsilon-carboxymethyllysine as a degradation product of fructoselysine in glycated protein. *J. Biol. Chem.* **26,** 4889–4894.
20. Prabhakaram, M., Smith, J. B., and Ortwerth, B. J. (1996) Rapid assessment of early glycation products by mass spectrometry. *Biochem. Mol. Biol. Int.* **40,** 315–325.
21. Smith, J. B., Sun, Y., Smith, D. L., and Green, B. (1992) Identification of the post-translational modifications of bovine lens alpha B-crystallins by mass spectrometry. *Protein Sci.* **1,** 601–608.
22. Lapolla, A., Baldo, L., Aronica, R., Gerhardinger, C., Fedele, D., Elli, G., Seraglia, R., Catinella, S., and Traldi, P. (1994) Matrix-assisted laser desorption ionization mass spectrometric studies on protein glycation 2. The reaction of ribonuclease with hexoses. *Biol. Mass. Spectr.* **23,** 241–248.
23. Lapolla, A., Fedele, D., Seraglia, R., Catinella, S., and Traldi, P. (1994) Matrix-assisted laser desorption/ionization capabilities in study of non-enzymatic protein glycation. *Rapid Commn. Mass Spectr.* **8,** 645–52.
24. Prabhakaram, M. and Ortwerth, B. J. (1994) Determination of glycation crosslinking by the sugar-dependent incorporation of [^{14}C]lysine into protein. *Anal. Biochem.* **216,** 305–312.
25. Lopez, M. G. and Feather, M. S. (1992) The production of threose as a degradation product from L-ascorbic acid. *J. Carbohydr. Chem.* **11,** 799–806.
26. Mans, R. J. and Novelli, G. D. (1961) Measurement of the incorporation of radioactive amino acids into protein by a filter-paper disk method. *Arch. Biochem. Biophys.* **94,** 48–53.
27. Lee, K., Mossine, V., and Ortwerth, B. J. (1998) The relative ability of glucose and ascorbate to glycate and crosslink lens proteins in vivo. *Exp. Eye Res.* **67,** 95–104.

Index